W9-BID-377

25686   70326038
Roberto R Mite
Product Mgr
Adcomm Inc
89 Leuning St
South Hackensack NJ 07606-1335

Advanced Techniques in RF Power Amplifier Design

For a listing of recent titles in the *Artech House Microwave Library*, turn to the back of this book.

Advanced Techniques in RF Power Amplifier Design

Steve C. Cripps

Artech House
Boston • London
www.artechhouse.com

Library of Congress Cataloging-in-Publication Data
Cripps, Steve C.
   Advanced techniques in RF power amplifier design / Steve Cripps.
      p.   cm. — (Artech House microwave library)
   Includes bibliographical references and index.
   ISBN 1-58053-282-9 (alk. paper)
   1. Power amplifiers.   2. Amplifiers, Radio frequency.   I. Title.   II. Series.
TK7871.58.P6 C72   2002
621.384'12—dc21   2002016427

British Library Cataloguing in Publication Data
Cripps, Steve C.
   Advanced techniques in RF power amplifier design. — (Artech House
   microwave library)
   1. Power amplifiers—Design   2. Amplifiers, Radio frequency
   I. Title
   621.3'8412

   ISBN 1-58053-282-9

Cover design by Gary Ragaglia

International Standard Book Number: 1-58053-282-9
Library of Congress Catalog Card Number: 2002016427

10 9 8 7 6 5 4 3 2 1

Contents

	Preface	*xi*

	Acknowledgments	*xv*

1	**Class AB Amplifiers**	**1**
1.1	Introduction	1
1.2	Classical Class AB Modes	2
1.3	Class AB: A Different Perspective	9
1.4	RF Bipolars: *Vive La Difference*	16
1.4.1	A Basic RF BJT Model	16
1.5	On Sweet Spots and IM Glitches	29
1.6	Conclusions	31
	Reference	32

2	**Doherty and Chireix**	**33**
2.1	Introduction	33
2.2	The Doherty PA	34
2.2.1	Introduction and Formulation	34

2.2.2	The Classical Doherty Configuration	37
2.2.3	Variations on the Classical Configuration	42
2.2.4	Peaking Amplifier Configurations	49
2.2.5	Doherty PA Matching Topologies	52
2.2.6	The Multiple Doherty PA	56
2.2.7	Doherty PA Conclusions	56
2.3	The Chireix Outphasing PA	58
2.3.1	Introduction and Formulation	58
2.3.2	Discussion, Analysis, and Simulation	62
2.3.3	Variations	69
2.3.4	Chireix: Conclusions	71
	References	72
3	**Some Topics in PA Nonlinearity**	**73**
3.1	Introduction	73
3.2	A Problem, a Solution, and Problems with the Solution	74
3.3	Power Series, Volterra Series, and Polynomials	77
3.4	Two-Carrier Characterization	89
3.5	Memory and IM Asymmetry in RFPAs	94
3.6	PAs and Peak-to-Average Ratios	105
3.7	Conclusions	109
	References	110
4	**Feedback Techniques**	**111**
4.1	Introduction	111
4.2	Feedback Techniques	114
4.3	Amplitude Envelope Feedback: Configuration and Analysis	121
4.4	Vector Envelope Feedback	137

4.5	Low Latency PA Design	140
4.6	Variations	146
4.7	Conclusions	150
	References	151
5	**Predistortion Techniques**	**153**
5.1	Introduction	153
5.2	Third-Degree PA: Predistortion Analysis	156
5.3	PD Characteristic for General PA Model	163
5.4	Practical Realization of the Predistorter Function: Introduction	177
5.5	Analog Predistorters	179
5.6	DSP Predistortion	187
5.7	Conclusions	194
	References	195
6	**Feedforward Power Amplifiers**	**197**
6.1	Introduction	197
6.2	The Feedforward Loop	198
6.3	AM-PM Correction in the Feedforward Loop	204
6.4	Error Insertion Coupling	208
6.5	Third-Degree Analysis of the Generalized Feedforward (FFW) Loop	216
6.5.1	Formulation and Analysis	216
6.5.2	Tracking Errors	222
6.5.3	Compression Adjustment	224
6.5.4	Third-Degree Analysis: Conclusions	228
6.6	Feedforward Loop Simulation	229
6.6.1	Effect of AM-PM in Main PA	231
6.6.2	Gain Compression Adjustment	233

6.6.3   EPR Change   233

6.6.4   Gain and Phase Tracking   235

6.6.5   Multicarrier Simulation   235

6.7   Feedforward Loop Efficiency Considerations   236

6.8   Adaption and Correction: Closing the Loop   241

6.9   Variations   249

6.9.1   The Double Feedforward Loop   249

6.9.2   A Feedforward-Enhanced Power Combiner   252

6.9.3   "Budget" FFW Systems   253

6.10   Conclusions   255

         References   256

7   Microwave Power Amplifiers   257

7.1   Introduction   257

7.2   Broadband Microwave Power Amplifier Design   259

7.2.1   Introduction   259

7.2.2   Broadband Matching Using Network Synthesis   259

7.2.3   Balanced Amplifiers   270

7.2.4   Broadband Power Amplifier Design Issues   273

7.3   Microwave Circuits and MIC Techniques   280

7.3.1   Introduction   280

7.3.2   Substrate and Heatsink Materials   281

7.3.3   MIC Components and Structures   283

7.4   PA Design Using Prematched Modules   287

7.4.1   Introduction   287

7.4.2   Matching Issues for IMTs   287

7.4.3   Biasing Issues for IMTs   290

7.4.4   Power Combining of IMTs   291

7.5   Distributed Amplifiers   293

7.6   Conclusions   296

         References   297

Appendix **299**

Notes 299

Selected Bibliography 300

Glossary **301**

About the Author **305**

Index **307**

Preface

First of all, I should explain the title of this book. This is not really an advanced book, but does cover topics which are generally of a less elementary, or tutorial, nature than my first book. In *RF Power Amplifiers for Wireless Communications* (*RFPA*, also published by Artech House), my overriding goal was to present the subject material in a manner that is analytical but hopefully still readable. Engineering literature has always bothered me. Both books and technical journals seem to present everything couched in high-level mathematics which the majority of practicing engineers can't understand, or at best don't have the time or inclination to decipher. There seems to be an "emperor's new clothes" situation about it all; make the subject as difficult as possible, frequently much *more* difficult than it needs to be, and you are almost assured of pious nods of approval in higher places. Lower places, where most of us operate, offer a less reverent reception, but seem to accept the situation nevertheless. The odd paper maybe progresses as far as the fax machine or the filing cabinet, but then rests in peace and gathers dust. I am encouraged, and very grateful, for the positive response to my modest but radical attempts to change this situation in *RFPA*, and proceed with this volume in very much the same spirit.

The intention is that this book can be regarded as a sequel to the first, but can also be read in isolation. It would be appropriate, therefore, to restate briefly some of the philosophy and goals that carry over. First and foremost, the spirit of *a priori* design methods remains paramount. Simulation tools are getting better all the time, but the advantages of performing symbolic analysis using simplified models before engaging in number crunching are still persuasive. Indeed, not being a particularly advanced mathematician myself helps in that I feel it is easier to present readable symbolic analysis, rather than the seemingly statutory unreadable stuff which abounds in the more "learned" literature.

This book covers some new topics that barely got a mention in *RFPA*, and takes some others which, although covered in the first book, deserve more detailed treatment. Bipolar junction transistors (BJTs), for instance, were barely mentioned in *RFPA* and with the advent of heterojunction bipolar transistors (HBTs) and Silicon Germanium (SiGe) technology, things are hotting up again for the BJT (pun somewhat intended). Chapter 1 revisits the basics of Class AB operation, but with greater emphasis on BJT applications. Microwave power amplifier applications also get a chapter of their own, Chapter 7, in which an attempt is made to pull together some of the techniques which were developed in the 1970s and 1980s for higher frequency (> 2 GHz) broadband amplifier design, and which could easily come back on stream again as wireless communications run out of bandwidth at the low end of the microwave spectrum.

Other chapters pick up on some of the things that were introduced in *RFPA*, but not wrung out to some readers' satisfaction. For all my enthusiastic promotion on the Doherty and Chireix techniques in *RFPA*, these admirable inventions still appear to remain firmly rooted in history and vacuum tube technology. So Chapter 2 revisits these topics regarding especially their potential role in the modern scene. This includes a more generalized analysis of the Doherty PA and some simulation results on practical implementations. The Chireix method continues to be a talking point and not much else; Chapter 2 attempts to dig a little deeper into why this should be.

Chapter 3 takes up the theme of nonlinear effects in PAs, with a particular emphasis on the problem of asymmetrical IM distortion, its causes and possible remedies. This chapter also has some tutorial material on the behavioral modeling of PAs and attempts to show that there is still much useful life left in polynomials and Volterra series modeling methods. In particular, the use of dynamic, rather than static, characterization methods are discussed in relation to memory effects.

Linearization is, of course, an inescapable aspect of the modern PA scene. Equally inescapable, it seems, is the growing impact of digital signal processing (DSP) techniques in this area. Chapters 4, 5, and 6 cover the three main linearization topics of feedback, predistortion, and feedforward. In these chapters I find myself up against a burgeoning and already voluminous literature, and also topics of current research which are engaging many thousands of engineers all over the world. I have therefore not attempted to cover each topic in an exhaustive manner. I have basically applied my statutory methods of symbolic analysis using simplified models, and I believe come up with some useful results and observations. Most of the material in Chapters 4, 5, and 6 represents, I believe, at least a different angle on the subject and as such is hopefully complementary to existing published accounts. As a PA designer myself, one aspect which I have attempted to emphasize throughout these three chapters is the need to redefine, and even to rethink, the design of a PA which is to be used in any linearization system. This is an aspect of the business which I feel has not been adequately considered, for the logistical reason that PA linearization and PA design are frequently done by different groups of people or different organizations.

I must make some comments on the patent situation in this field. This book deals with a technology that has been the focus for much patent activity over the years, and especially within the last decade or so. It seems that patents are now being issued not just for specific implementations, or variations of well-established techniques, but in some cases for the well-established techniques themselves. This is a big headache for any company wishing to enter into the wireless communications PA business, and poses a problem for an author as well. To generate a comprehensive, exhaustive list of relevant patents would be a task comparable to writing a whole new book. Indeed, perhaps it is a book someone should write. My policy has been to refrain completely from citing individual patents as references, other than one or two historical ones. This avoids any conflict where opposing factions may be claiming priority and only one gets the citation. It does not, however, avoid the possibility that I may be describing, or proposing, something that has been patented sometime, by somebody. I have tried in general throughout this book to make the reader aware of the need to perform patent searches in this business if a commercial product is being contemplated. I have also made specific comments about the likelihood of patents in certain focused areas. In general, I have included a few ideas of my own in most chapters (usually under the subheading "Variations"). These suggestions are my own independent ideas and do not represent any commercial products of which I am aware. They have not, however, been subjected to any patent search.

Unlike the previous book, which had substantial continuity from chapter to chapter, this book treats the numerous topics in a manner which does not always fall into a seamless narrative. Such is the nature of "more advanced topics." This is, primarily, a *theoretical* book; for the most part I am analyzing how things work, and developing *a priori* methods for designing them. It is not a step-by-step guide on how to build RF power amplifiers, advanced or otherwise. I believe that I am addressing topics which RF designers, and especially those involved with RF power amplifiers, *talk* about a lot amongst themselves. I therefore make no apologies for using more words, and fewer equations, than in a conventional technical book.

Acknowledgments

This book represents, in a chronological sense, a time period in my own technical career which extends approximately back to my return from California, to the greener and damper Blackdown Hills of Somerset in England, where I have managed to keep working in the RFPA business, thanks mainly to the sponsorship of several clients. Also, I continue to find the intelligent questions of my PA design training course attendees a great stimulus for keeping on top of a rapidly developing technical area. So I must also acknowledge this time, having omitted to mention them last time, the ongoing sponsorship I have from CEI Europe, who continue to offer a first-class service and organization for training RF and communications professionals in Europe.

Steve C. Cripps
Somerset, England
May 2002

1

Class AB Amplifiers

1.1 Introduction

The Class AB mode has been a focus for several generations of power amplifier designers, and for good reasons. It is a classical compromise, offering
higher efficiency and cooler heatsinks than the linear and well-behaved Class
A mode, but incurring some increased nonlinear effects which can be tolerated, or even avoided, in some applications. The main goal in this chapter is
to invite PA designers and device technologists to break out of the classical
Class AB tunnel vision which seems to afflict a large proportion of their
numbers. For too long, we have been assuming that our radio frequency
power amplifier (RFPA) transistors obediently conduct precisely truncated
sinewaves when the quiescent bias is reduced below the Class A point,
regardless of the fact that an RF power device will typically have nowhere
near the switching speed to perform the task with the assumed precision. The
irony of this is that the revered classical theory, summarized in Section 1.2,
actually makes some dire predictions about the linearity of a device operated
in this manner, and it is the sharpness of the cutoff or truncation process that
causes some of the damage. Decades of practical experience with RFPAs of
all kinds have shown that things generally work out better than the theory
predicts, as far as linearity is concerned, which has relegated the credibility of
the theory. For this reason, and others, flagrantly empirical methods are still
used to design RFPAs, in defiance of modern trends.

Section 1.3 attempts to reconcile some of the apparent conflict between
observation and theory, showing that an ideal device with a realizable

characteristic can be prescribed to allow linear operation along with near classical efficiency. Section 1.4 discusses the RF bipolar and its radically different formulation for reaching the same goal of linearity combined with high-efficiency operation. The RF bipolar emerges from this analysis, taking full account of the discussion in Section 1.3, in a surprisingly favorable light. Section 1.5 returns to the field effect transistor (FET) as a Class AB device, and the extent to which existing devices can fortuitously exhibit some of the linearization possibilities discussed in Section 1.3.

1.2   Classical Class AB Modes

This analysis should need no introduction, and what follows is largely a summary of a more detailed treatment in *RFPA*, but with some extensions into the possibilities offered by dynamically varying RF loads. Figure 1.1 shows an idealized RF device, having a linear transconductive region terminated by a sharply defined cutoff point. The device is assumed to be entirely transconductive, that is to say, the output current has no dependency on the output voltage provided this voltage is maintained above the turn-on, or "knee" value, V_k. The analysis will further make the approximation that V_k is

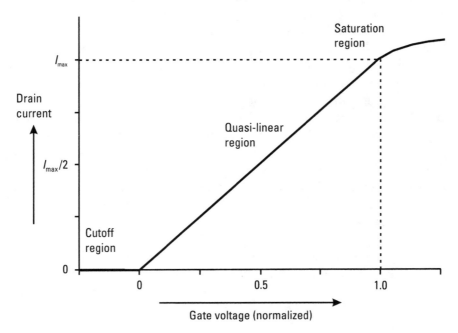

Figure 1.1 Ideal transconductive device transfer characteristic.

negligible in comparison to the dc supply voltage, in other words, zero. This approximation is conspicuously unreal, and needs immediate addressing if the voltage is anything other than sinusoidal, but is commonplace in elementary textbooks. Figure 1.2 shows the classical circuit schematic for Class AB operation. The device is biased to a quiescent point which is somewhere in the region between the cutoff point and the Class A bias point. The input drive level is adjusted so that the current swings between zero and I_{max}, I_{max}

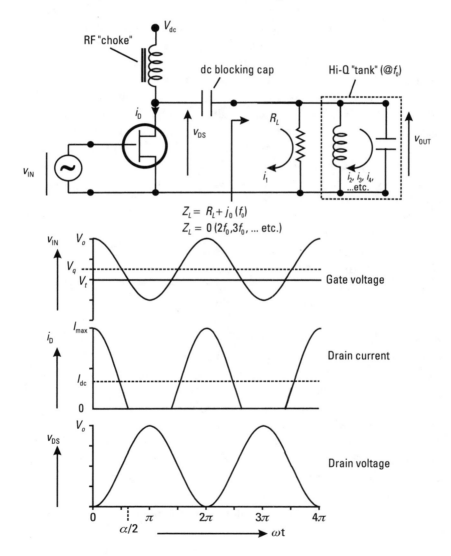

Figure 1.2 Class AB amplifier: schematic and waveforms.

being a predetermined maximum useable current, based on saturation or thermal restrictions.

The resulting current waveforms take the form of asymmetrically truncated sinewaves, the zero current region corresponding to the swings of input voltage below the cutoff point. These current waveforms clearly have high harmonic content. The key circuit element in a Class AB amplifier is the harmonic short placed across the device which prevents any harmonic voltage from being generated at the output. Such a circuit element could be realized, as shown in Figure 1.2, using a parallel shunt resonator having a resonant frequency at the fundamental. In principle the capacitor could have an arbitrarily high value, sufficient to short out all harmonic current whilst allowing the fundamental component only to flow into the resistive load. So the final output voltage will approximate to a sinewave whose amplitude will be a function of the drive level and the chosen value of the load resistor. In practice the load resistor value will be chosen such that at the maximum anticipated drive level, the voltage swing will use the full available range, approximated in this case to an amplitude equal to the dc supply. For the purposes of this analysis, the maximum drive level will be assumed to be that level which causes a peak current of I_{max}.

Some simple Fourier analysis [1] shows that the efficiency, defined here as the RF output divided by the dc supply, increases sharply as the quiescent bias level is reduced, and the so-called conduction angle drops (Figure 1.3). Not only does this apply to the efficiency at the designated maximum drive level, but the efficiency in the "backed-off" drive condition also increases, especially in relation to the Class A values (Figure 1.4). What is less familiar is the plot of linearity in the Class AB region, shown in Figure 1.5. The process of sharp truncation of the input sinusoidal signal unfortunately generates some less desirable effects; odd degree distortion is part of the process and gain compression is clearly visible anywhere in the Class AB region. This gain compression comes from a different, and additional, source than the gain compression encountered when a Class A amplifier, for example, is driven into saturation. Saturation effects are primarily caused by the clipping of the RF voltage on the supply rails. The class AB nonlinearity in Figure 1.5 represents an additional cause of distortion which will be evident at drive levels much lower than those required to cause voltage clipping. This form of distortion is particularly undesirable in RF communications applications, where signals have amplitude modulation and stringent specifications on spectral spreading.

The Class B condition, corresponding to a zero level of quiescent bias, is worthy of special comment. This case corresponds to a current waveform

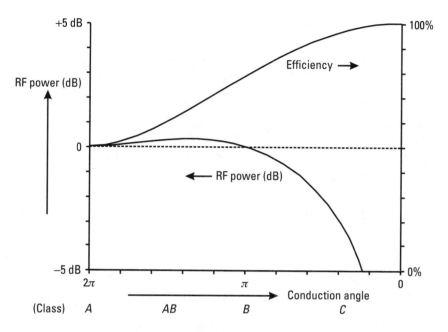

Figure 1.3 Reduced conduction angle modes, power and efficiency at maximum drive level.

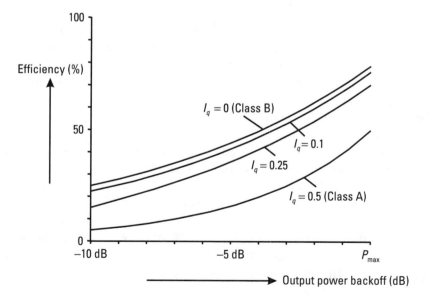

Figure 1.4 Efficiency as a function of input drive backoff (PBO) and Class AB "quiescent" current (I_q) setting.

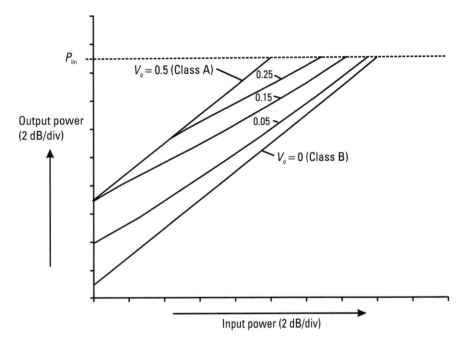

Figure 1.5 Class AB gain characteristics.

which, within the current set of idealizing assumptions, is a perfectly half-wave rectified sinewave. Such a waveform contains only even harmonics, and in the absence of damaging odd degree effects, the backed-off response in Figure 1.5 shows a return to linear amplification. In practice, such a desirable situation is substantially spoiled by the quirky, or at best unpredictable, behavior of a given device so close to its cutoff point. It is frequently found, usually empirically, that a bias point can be located some way short of the cutoff point where linearity and efficiency have a quite well-defined optimum. Such "sweet spots" are part of the folklore of RFPA design, and some aspects of this subject will be discussed in more detail in Section 1.3.

One additional aspect of Class AB operation which requires further consideration is the issue of drive level and power gain. It is clear from Figure 1.2 that as the quiescent bias point is moved further towards the cutoff point, a correspondingly higher drive voltage is required in order to maintain a peak current of I_{max}. In many cases, especially in higher RF or microwave applications, the gain from a PA output stage is a hard-earned and critical element in the overall system efficiency and cost. In moving the bias point from the Class A ($I_{max}/2$) point to the Class B (zero bias) point, an increase of drive level of a factor of two is required in order to maintain a peak current of I_{max}. This corresponds to an increase of 6 dB in drive level, and this is equally a reduction in the power gain of the device. It is common practice to compromise this problem by operating RF power devices at some lower level than I_{max}, in order to preserve efficient operation at higher power gain. The process is illustrated in Figure 1.6 for a Class B condition. If, for example, the drive level is increased only 3 dB from the Class A level, the current peaks, in a zero bias condition, will only reach $I_{max}/\sqrt{2}$. This reduction in maximum linear power can be offset by increasing the value of the fundamental load resistor by the same ratio of $\sqrt{2}$. The result, shown in the second set of waveforms in Figure 1.6, shows only a 1.5-dB reduction in power at the available maximum drive, compared to the fully driven case. Significantly, however, the efficiency in the "underdriven" case returns to the original value of 78.5%.

This concept of "underdrive" can be extended to more general Class AB cases, although in the Class AB region the efficiency will not return to the

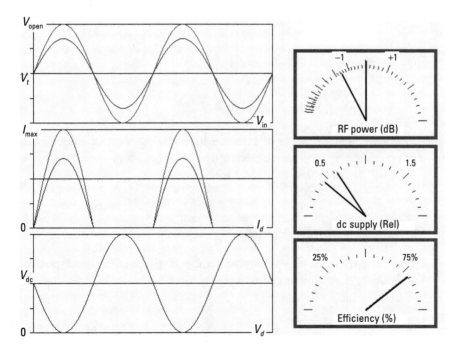

Figure 1.6 Class B operation: "Fully driven" condition gives the same power as Class A (0 dB) but requires a 6-dB higher input drive. "Underdriven" condition (3-dB underdrive case shown) can still give full Class B efficiency if load resistor is adjusted to give maximum voltage swing.

fully driven value due to the effective increase of conduction angle caused by drive reduction. Another extension of the concept is to consider the possibility of an RF load resistor whose value changes dynamically with the input signal level. Such an arrangement forms one element of the Doherty PA which will be discussed in more detail in Chapter 2. It is, however, worthy of analysis in its own right, on the understanding that it does not at this stage constitute a full Doherty implementation.

Suppose that, by some means or other, the value of the load resistor is caused to vary in inverse proportion to the signal amplitude v_s,

$$R_L = R_o / v_s$$

so that as the fundamental component of current, I_1, increases from zero to $I_{max}/2$, the fundamental output voltage amplitude remains constant at

$$v_o = \left(\frac{R_o}{v_s} \right) \left(\frac{I_{max}}{2} \right) v_s = V_{dc}$$

where

$$I_1 = \left(\frac{I_{max}}{2} \right) v_s$$

making the usual assumption of a perfect harmonic short, and a device knee voltage which is negligible compared to the dc supply, V_{dc}.

The fundamental output power is therefore

$$P_o = \left(\frac{V_{dc}}{2} \right) \left(\frac{I_{max}}{2} \right) v_s \qquad (1.1a)$$

which is unusual in that the output power is now proportional to input *voltage* amplitude, rather than input power.

The efficiency is given by

$$\eta = \left(\frac{\pi}{V_{dc} v_s I_{max}} \right) \left(\frac{V_{dc}}{2} \right) \left(\frac{I_{max}}{2} \right) v_s = \frac{\pi}{4} \qquad (1.1b)$$

Equation (1.1b) is an interesting result, the efficiency being independent of the signal drive level. Given that the two central issues in modern PA design are firstly the rapid drop in efficiency as a modulated signal drops to low envelope amplitudes, and secondly the need to control power over a wide dynamic range [for example, in code division multiple access (CDMA) systems], this configuration appears to fulfill both goals handsomely. There are also, of course, two immediate problems; the device is a nonlinear amplifier having a square-root characteristic, and we have so far ignored the practical issue of how such a dynamic load variation could be realized in practice.

As will be discussed in Chapter 2, the realization of an RF power amplifying system capable of performing the feat of linear high efficiency amplification over a wide dynamic signal range has been something of a "Holy Grail" of RFPA research for over half a century. Both the Doherty and Chireix techniques (Chapter 2) are candidates, but also generate a collection of additional, mainly negative, side issues. The fundamental principle remains sound, and is an intriguing goal for further innovative research.

1.3 Class AB: A Different Perspective

The idealized analysis of Class AB modes summarized in Section 1.2 raises a number of issues for those who have experience in using such amplifiers in practice. Most prominently, the assumption of a linear transconductive device is an idealization that is unsatisfactory for just about any variety of RF device in current use, whether it be an FET or bipolar junction transistor (BJT). It seems that in practice the use of an imperfect device can fortuitously reduce the nonlinearities caused by the use of reduced angle operation. This section explores this extension to the theory and comes up with some proposals concerning the manner in which RF power transistors should be designed and specified. In essence, devices with substantial, but correctly orientated, nonlinear characteristics are required to make power amplifiers having the best tradeoff between efficiency and linearity. The process of defining such devices involves some basic mathematical analysis and flagrantly ignores, for the present purposes, the technological issues involved in putting the results into practice. This is a necessary and informative starting point.

The analysis in Section 1.2 showed that an ideal transconductive device, biased precisely at its cutoff point, gives an optimum linear amplifier, having high efficiency and a characteristic which contains even, but not odd, degree nonlinearity. This is the classical Class B amplifier. It has already been commented that in practice, true "zero-bias" operation usually yields unsatisfactory performance, especially at well backed-off drive levels where the device will typically display a collapse of small signal gain. Even a device with an ideal, sharp characteristic does not stand up so well under closer scrutiny. Figure 1.7 shows the third-order intermodulation (IM3) response for an ideal device biased a small way either side of the ideal cutoff, or Class B, point. Clearly, the favorable theoretical linearity of a Class B amplifier is a very sensitive function of the bias point, and indicates a critical yield issue in a practical situation. In this respect, the ideally linear transconductive device may not be such an attractive choice for linear, high-efficiency applications as it may at first appear, and some alternatives are worth considering.

An initial assumption used in this analysis is that RF transistors have characteristics which are curves, as opposed to straight lines; attempts to make a device having the ideal "dogleg" transconductive characteristic shown in Figure 1.1 will be shown to be misdirected. A useful starting point is a square-law transconductive device, shown in Figure 1.8. In all of the following analyses, the device characteristic will be normalized such that the maximum current, I_{max}, is unity and corresponds to a device input voltage of unity. The zero current point will correspond to an input also normalized to zero; the input voltage, unlike for conventional Class AB analysis, will not be allowed to drop below the normalized zero point. So the square-law characteristic is, simply,

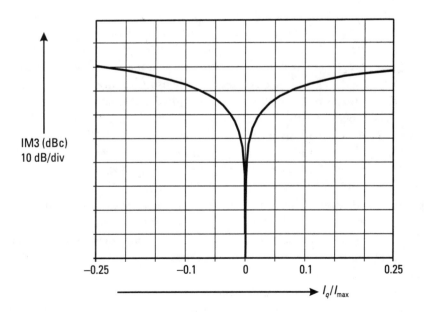

Figure 1.7 IM3 response of ideal transconductive device in vicinity of Class B quiescent bias point.

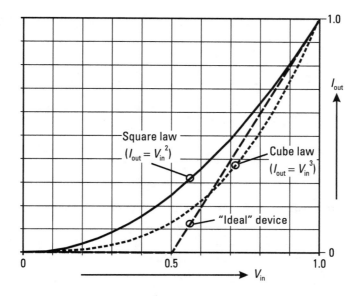

Figure 1.8 Square-law and cube-law device characteristics, compared to ideal device using linear and cutoff regions. (Note changed normalization for ideal device.)

$$i_o = v_i^2$$

and for maximum current swing under sinusoidal excitation, the quiescent bias point will be set to $v_i = 0.5$, and the input signal will be to $v_i = v_s \cos\theta$, with v_s varying between zero and a maximum value of 0.5. So the output current for this device will be given by

$$i_o = \left(\tfrac{1}{2} + v_s \cos\theta\right)^2, \quad 0 < v_s < \tfrac{1}{2} \tag{1.2}$$

so that

$$i_o = \left(\tfrac{1}{4} + v_s \cos\theta + v_s^2 \cos^2\theta\right),$$
$$= \left\{\left(\tfrac{1}{4} + \tfrac{1}{2}v_s^2\right) + v_s \cos\theta + \tfrac{1}{2}v_s^2 \cos 2\theta\right\}$$

Assuming that the output matching network presents a short circuit at all harmonics of θ, the fundamental output voltage amplitude is a *linear* function of the input level, v_s, despite the square-law device characteristic.

Compared to a device with an ideal linear characteristic in Class A operation, where

$$i_o = \left(\tfrac{1}{2} + v_s \cos\theta\right)$$

the square-law device has the same fundamental output amplitude, but a dc component reduced by a factor of

$$\frac{\tfrac{1}{4} + \tfrac{1}{2}v_s^{\,2}}{\tfrac{1}{2}} = \tfrac{1}{2}\left(1 + 2v_s^{\,2}\right)$$

which improves the output efficiency over the corresponding linear Class A value. So at the maximum drive level of $v_s = 0.5$, the efficiency of the square-law device is 2/3 or 66.7%. This improvement in efficiency is obtained simultaneously with perfectly linear amplification.

A cube-law device (Figure 1.8), on the other hand, gives substantial improvement in efficiency at the expense of linearity,

$$i_o = \left(\tfrac{1}{2} + v_s \cos\theta\right)^3$$
$$= \tfrac{1}{8}\left(1 + 6v_s^{\,2}\right) + \tfrac{3}{4}\left(v_s + v_s^{\,3}\right)\cos\theta + \tfrac{3}{4}\left(v_s^{\,2}\right)\cos 2\theta + \tfrac{1}{4}\left(v_s^{\,3}\right)\cos 3\theta$$

(1.3)

showing an increased fundamental component compared to the linear case, and a reduced dc component. The output efficiency,

$$\eta = \frac{\left(\tfrac{1}{2}\right)v_s\left(\tfrac{3}{4}\right)\left(1 + v_s^{\,2}\right)}{\left(\tfrac{1}{8}\right)\left(1 + 6v_s^{\,2}\right)} = \frac{3v_s\left(1 + v_s^{\,2}\right)}{1 + 6v_s^{\,2}}$$

is now 3/4, or 75%, at maximum current swing ($v_s = 0.5$), but at this drive level the device displays 1.9 dB of gain expansion, leading to substantial generation of third-degree nonlinearities.

It is therefore apparent that to create a device which has optimum efficiency and perfect linearity, it is necessary to tailor the transfer characteristic to generate only even powers of the cosine input signal. Unfortunately, this is not as simple as creating a power transfer characteristic having a higher even order power,

$$i_o = \left(\tfrac{1}{2} + v_s \cos\theta\right)^4$$

which will contain both even and odd powers of the cosine signal.

The necessary characteristic can be determined by finding suitable coefficients of the even harmonic series,

$$i_o = k_o + \cos\theta + k_2 \cos 2\theta + k_4 \cos 4\theta + k_6 \cos 6\theta + \ \ldots \ + k_{2n} \cos 2n\theta$$

normalized such that $0 < i_0 < 1$.

The goal here is to find a set of coefficients which generates a waveform having the same peak-to-peak swing, from zero to unity, but which has decreasing mean value as successive even harmonic components are added. The optimum case will be a situation where the negative half cycles have a maximally flat characteristic at their minima; this corresponds to values of the k_n coefficients determined by setting successive derivatives of the function

$$f(\theta) = \cos\theta + k_2 \cos 2\theta + k_4 \cos 4\theta + k_6 \cos 6\theta + \ \ldots \ + k_{2n} \cos 2n\theta$$

equal to zero at $\theta = \pi$. This generates a set of simultaneous equations for the k_n values. Some of the resulting waveforms are shown in Figure 1.9, from which it is clear that as more even harmonics are added, the resulting waveforms more closely approach an ideal Class B form. Table 1.1 shows the values for a number of values of n, along with the efficiency, which improves for higher values of n due to the decreasing values of mean current, k_0.

Clearly, Table 1.1 shows that a useful increase in efficiency can be obtained for a few values of n above the simple square-law case of $n = 1$. Conversely, the large n values required to approach the Class B condition are unlikely to be realized in practice due to the limited switching speed of a typical RF transistor.

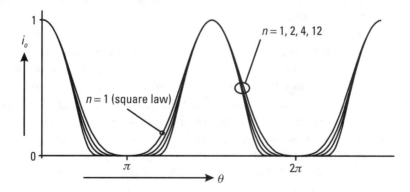

Figure 1.9 Current waveforms having "maximally flat" even harmonic components (n factor indicates the number of even harmonics).

Table 1.1
Even Harmonic Efficiency Enhancement

n	k_0	k_2	k_4	k_6	k_8	$\eta(\%)$
1	0.75	0.25				66.7
2	0.703	0.3125	−0.0156			70.3
3	0.6835	0.3428	−0.0273	0.00195		73.1
4	0.673	0.3589	−0.0359	0.00439	−0.0003	74.3
8	0.656					76.2
12	0.6495					77.0

It is a simple matter to convert the desired current waveforms shown in Figure 1.9 into corresponding transfer characteristics, assuming a sinusoidal voltage drive. These corresponding nonlinear transconductances are shown in Figure 1.10. It seems that an unfamiliar device characteristic emerges from this simple analysis, which displays efficiency in the mid-70% region and has only even order nonlinearities. It has a much slower turn-on characteristic than the classical FET dogleg, and resembles a bipolar junction transistor (BJT), rather than an FET in its general appearance. Figure 1.10 also shows that the desired family of linear, highly efficient characteristics fall into a well-defined zone. The boundaries of the zone are formed by the square-law characteristic, and the classical Class B dogleg. It is interesting to plot some other characteristics on the same chart, as shown in Figure 1.11. The characteristics which have inherent odd degree nonlinearities always cross over the boundary formed by the dogleg. The linearity zone, thus defined, would appear to be a viable and realistic target for device development.

The chart of Figure 1.11 has some interesting implications for the future of RF power bipolars. This will be further discussed in Section 1.4. FETs, however, do not fare so well in this analysis. An FET will usually display a closer approximation to a dogleg characteristic; this is a natural outcome of their *modus operandi*, coupled with some misdirected beliefs on the part of manufacturers as to what constitutes a "good" device characteristic. It could be reasonably argued that a typical FET characteristic has the appearance of one of the higher *n*-value curves in Figure 1.10, having linear transconductance with a short turn-on region. Such a device would, within the idealized boundaries of the present analysis, still comply with the

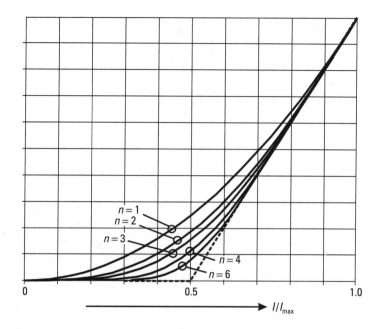

Figure 1.10 Device characteristics "tailored" to give current waveforms having only even harmonics, as shown in Figure 1.9, for sinusoidal voltage input. Conventional Class B using linear device is shown dotted.

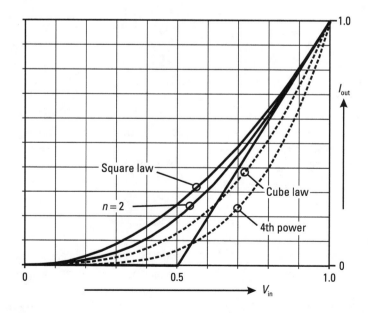

Figure 1.11 Linearity "zone" (solid line) for Class AB device characteristics.

requirements of even degree nonlinearity and higher efficiency than the lower n-value curves. The problem with this kind of device lies in the precision of the quiescent bias setting, which leads to more general issues of processing yield. It is fair to speculate that the unfamiliar-looking $n = 4$ curve, for example, would be a more robust and reproducible device for linear power applications.

1.4   RF Bipolars: *Vive La Difference*

The idealized analysis of Class AB modes summarized in Section 1.1 has its roots in tube amplifier analysis and dates from the early part of the last century. The early era of RF semiconductors was dominated by a radically different kind of device, the bipolar transistor. More recently, the emergence of RF FET technologies, such as the Gallium Arsenide Metal Semiconductor Field Effect Transistor (GaAs MESFET) and Silicon Metal Oxide Semiconductor (Si MOS) transistor, has renewed the relevance of the older traditional analysis. Strangely, it seems that despite some obvious and fundamental physical differences in the manner of operation of BJTs, much of the conceptual framework and terminology of the traditional analysis seems to have been retained by the BJT RFPA community. This has required the application of some hand-waving arguments which seek to gloss over the major physical differences that still exist between BJT and FET device operation.

This section attempts to perform a complementary analysis of a BJT RF power amplifier, in the same spirit of device model simplicity as was used in analyzing the FET PAs in Section 1.1. Unfortunately, the exponential forward transfer characteristic of the BJT device will necessitate greater use of numerical, rather than purely analytical, methods. It will become clear that the BJT is a prime candidate for practical, and indeed often fortuitous, implementation of some of the theoretical results discussed in this section.

1.4.1   A Basic RF BJT Model

The model for the RF BJT which will be used is shown in Figure 1.12. This model incorporates the two essential textbook features of BJT operation: a base-emitter junction which has an exponential diode I-V characteristic, and a collector-emitter output current generator which supplies a multiplied replica of the current flowing in the base-emitter junction. As with the FET model, all parasitic elements are assumed to be either low enough to be ignored or to form part of the external matching networks which resonate

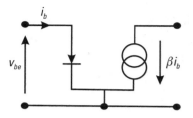

Figure 1.12 BJT model.

them out. Such assumptions are quite justifiable in the modern era where 30-GHz processes are frequently used to design PAs below 2 GHz. It is worth emphasizing, however, that the input and output parasitics, usually capacitances, can still be quite high even in processes which yield useful gain at millimeter-wave frequencies. The assumption of resonant matching networks for these parasitics will play an important role in the interpretation of some of the results.

The transfer characteristic for such a device is shown in Figure 1.13. Normalization of a BJT characteristic is not such a clear issue as for an FET. The maximum peak current, I_{max}, is usually well defined for an FET due to saturation. In the case of a BJT, the peak current is not so obviously linked to a physical saturation effect, and putting a value to I_{max} is a less well-defined process. We will, however, still continue to assume a predetermined value for

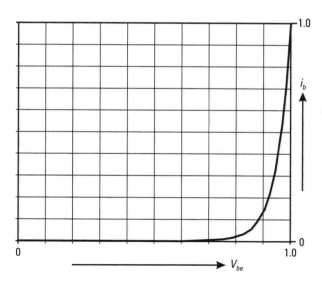

Figure 1.13 Normalized BJT transfer characteristic.

I_{max}, which will usually be based on thermal considerations for a BJT. This maximum current will be normalized to unity in the following analysis. There is an additional issue in the normalization process for a BJT, which is the steepness of the exponential base-emitter characteristic. For convenience, this will be modeled using values which give a typical p-n junction characteristic which turns on over approximately 10% of a normalized v_b range of 0 to 1. So

$$i_c = \beta i_b = \beta \left(\frac{e^{kv_b}}{e^k} \right)$$

where a k value of 7 and a normalized β value of unity give the characteristics shown in Figure 1.13. This closely resembles a typical BJT device except that the "on" voltage, where $i_c = 1$, is normalized to unity.[1]

Clearly, the immediate impression from Figure 1.13 is of a highly non-linear device. But this impression can be tempered by the realization that, unlike in the previous FET analysis, the voltage appearing across the base-emitter junction is no longer a linear mapping of the voltage appearing at the terminals of the RF generator; the input impedance of the RF BJT also displays highly nonlinear characteristics. It is the interaction of these two nonlinear effects which has to be unraveled, to gain a clear understanding of how one can possibly make linear RFPAs using such a device.

As usual, our elementary textbooks from younger days have a simple solution, and there is a tendency for this concept to be stretched, in later life, well beyond its original range of intended validity. Basically, if the device is fed from a voltage generator whose impedance, either internal to the generator or through the use of external circuit elements, is made sufficiently high compared to the junction resistance, then the base-emitter current approximates to a linear function of the generator voltage, which in turn appears in amplified form in the collector-emitter output circuit. This process is illustrated in Figure 1.14, where the effect of placing a series resistor on the base is shown for a wide range of normalized resistance values. The curves in Figure 1.14 are obtained by numerical solution of the equation

$$v_{in} = i_b R + 1 + \left(\frac{1}{k} \right) \log i_b$$

where R is normalized to 1Ω for normalized unity values of current and voltage.

---

1. It is also convenient to normalize β to unity, so that i_b and i_c are both normalized over a range of 0 to 1.

This simplification of BJT operation is the mainstay of most low-frequency analog BJT circuit design, but it has two important flaws in RFPA applications. The first problem concerns the optimum use of the available generator power. In RF power applications, power gain is usually precious and the device needs to be matched close to the point of maximum generator power utilization. This will typically imply a series resistance that is much lower in value than that required to realize the more extensive linearization effects shown in Figure 1.14. The second problem is that in order to make a Class AB type of amplifier, the output current waveform, and correspondingly the base-emitter current, has to be highly nonlinear. Thus the input series resistor has to be a "real" resistor, having linear broadband characteristics. This will not be the case if the series resistance is realized using conventional matching networks at the fundamental frequency. The fact that in a BJT the collector and base currents have to maintain a constant linear relationship is a crucial difference between FET and BJT amplifiers running in Class AB modes, and leads directly to the inconvenient prospect of harmonic impedance matching on the input, as well as the output.

Considering the Class A type operation initially, the need for a high Q input resonant matching network does enable the low frequency "current-gain" concept to be stretched into use. Figure 1.15 shows a situation closer to

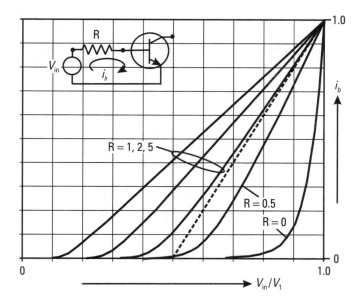

Figure 1.14 BJT transfer characteristics for varying base resistance. (Voltage scale normalized to V_1, the value of V_{in} required for $i_b = 1$ at each selected R value.)

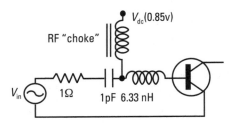

Figure 1.15 Schematic of BJT Class A RFPA, using high Q input resonator; circuit values shown for 2-GHz operation.

reality for the input circuit of a BJT RFPA. Provided that the resonator elements are chosen such that their individual reactances are large in comparison to the "on" junction resistance, the flywheel effect of the resonator will ensure that a sinusoidal current will flow into the base-emitter junction. For those who find the "flywheel" concept a little on the woolly side, the schematic of Figure 1.15 can be simulated using Spice; the resulting waveforms are shown in Figure 1.16. The high Q resonator, which in practice will incorporate an impedance step-down transformation from the 50-Ω generator source impedance, forces a sinusoidal current which in turn forces the base-emitter voltage to adopt a non-sinusoidal appearance. Provided that the base-emitter junction is supplied with a forward bias voltage that maintains a dc supply which is greater than the RF input current swing, fairly linear amplification, Class A style, will result. Such an amplifier could be designed quite successfully using the conventional constant current biasing arrangements used for small signal BJT amplifiers.

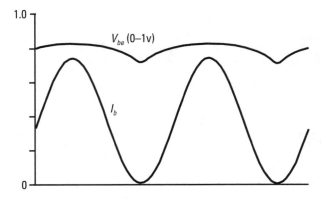

Figure 1.16 Spice simulated waveforms for Figure 1.15 schematic (input sinusoidal generator amplitude 0.5V).

Major problems will be encountered, however, if attempts are made to run this circuit configuration in a Class AB mode. Any effort to force a non-sinusoidal current in the base-emitter junction (such as by reducing the quiescent bias voltage) conflicts with the resonant properties of the input matching network, which will strongly reject harmonics through its high above-resonance impedance. In practice, the resonance of the input matching network may have only a moderate Q factor, and will allow some higher harmonic components to flow, giving some rather quirky approximations to Class AB or B operation. The harmonic current flow may also be aided by the BJT base-emitter junction capacitance, which will form part of the input-matching resonator. As discussed in *RFPA*, in connection with output harmonic shorts (see pp. 108–110), this leads to a curious irony in that higher frequency devices with lower parasitics can be harder to use at a given frequency from the harmonic trapping viewpoint. At 2 GHz, a typical Si BJT device will have an input which is dominated by a large junction capacitance. Although this makes the fundamental match a challenging design problem, it does have an upside in that higher harmonics will be effectively shorted out. A 40-GHz heterojunction bipolar transistor (HBT) device, however, will need assistance in the form of external harmonic circuitry.

Returning to the transfer characteristics plotted in Figure 1.14, it should be apparent that the intermediate values for series resistance give curves which are quite similar to those generated speculatively in Section 1.2 (see Figure 1.11). The BJT device appears to be a ready-made example of the novel principle that Class AB PAs can be more linear if the device has the right kind of nonlinearity in its transfer characteristic. This can be explored in a more quantitative fashion by taking the transfer characteristics in Figure 1.14 and subjecting the device to sinusoidal excitation. Figure 1.17(a) shows the circuit and defines the excitation. For convenience, the dc bias is assumed to be applied at the RF generator, and for the time being the input resistor is assumed to be a physical resistor, encompassing both the matched generator impedance and any additional resistance on the base. It should be noted that such a circuit configuration assumes that the resistor is a true resistor at *all relevant harmonic frequencies*. Although the input matching network can be assumed to transform the generator impedance to the R value at the fundamental, the circulating harmonic components of base current need to be presented with the same resistance value. One possible more practical configuration for realizing this requirement is shown in Figure 1.17(b). This initial analysis returns to the original assumption of ignoring the base-emitter capacitance; this approximation will be reviewed at a later stage.

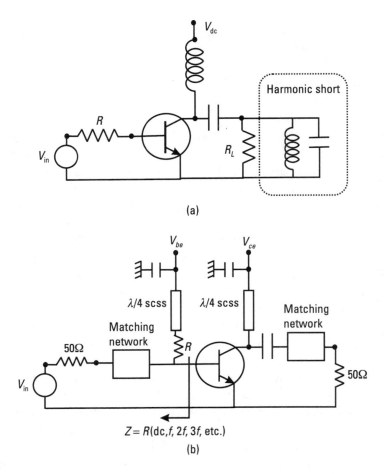

Figure 1.17 BJT Class AB circuits: (a) schematic for analysis; and "broadband" components and (b) possible practical implementation.

Unfortunately, the exponential base-emitter characteristic defies an analytical solution for the current flowing in the circuit of Figure 1.17(a), when using the instantaneous generator voltage as the independent input variable. An iteration routine has to be used at each point in the RF cycle to determine the junction current. A typical set of resulting waveforms is shown in Figure 1.18. These waveforms show three different cases of dc bias, resulting in corresponding quiescent current (I_q) values, for an input voltage swing chosen to give a stipulated maximum peak current (I_{max}) for the device. These current waveforms clearly resemble classical Class AB form, but are not precisely the same and have a complicated functional relationship with the

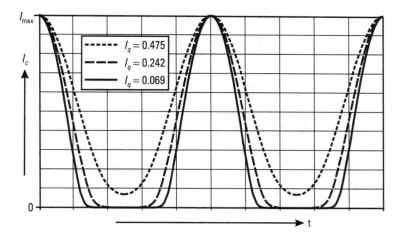

Figure 1.18 BJT "Class AB" current waveforms.

selected series resistance (input match), the bias point (which at this time is incorporated into the RF drive and has the same series resistance), and the RF drive level. Using these three essentially independent variables, we can explore the relationship between efficiency and linearity, at comparative output power levels.

The new variable factor in such an analysis is the choice of input resistance. Unlike the ideal FET analysis in Section 1.2, it now appears that the linearity of the amplifier, as well as its power gain, will have some important dependency on the selection of this circuit element. As always, it can be expected that some tradeoffs will be necessary. Power gain, efficiency, and linearity will all have different optimum values of input resistance. Figure 1.19 shows the gain compression and efficiency as a function of power back-off (PBO) for the three quiescent current settings in Figure 1.18. It is immediately clear that one case, corresponding to the "deep Class AB" quiescent current of 0.069 (6.9%), appears to give very linear power gain right up to the maximum peak current drive level, with a corresponding peak efficiency of just under 80%.

This desirable set of characteristics is closely coupled with the initial choice of normalized series resistance, $R = 0.5$. This value was selected on a qualitative comparison between the BJT transfer characteristics plotted in Figure 1.14 and the "linear zone" requirements suggested in Figure 1.11. It turns out that the only downside of this selection is that it represents a value substantially higher than the generator resistance required to achieve maximum power transfer to the device. Figure 1.20 shows a similar plot, but

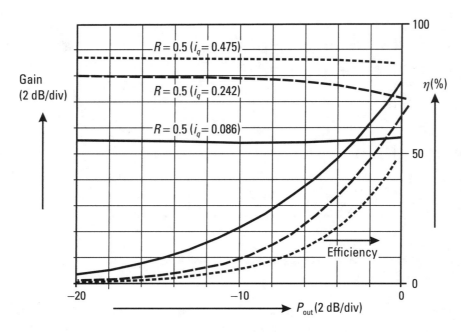

Figure 1.19 Gain compression and efficiency versus output power. (Figure 1.18 wave-forms correspond to maximum output power for each value of i_q.)

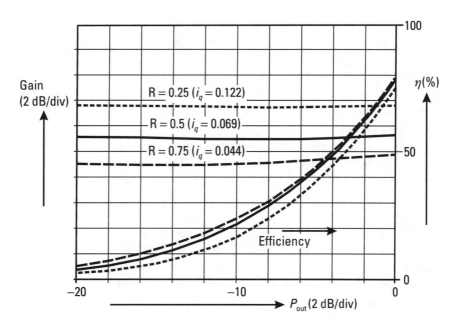

Figure 1.20 Gain and efficiency for different input resistance values.

showing three different values of series resistance ($R = 0.25$, 0.5, 0.75). In each case the quiescent bias point can be adjusted to give a linearity defined to be less than 0.5-dB gain variation over the 20-dB sweep range. The lower value of $R = 0.25$ shows a better input power match to the device, but requires a higher I_q setting in order to maintain the stipulated linearity; this lowers the efficiency. Higher R values result in similar good linearity at lower I_q settings and higher efficiency, but lower power gain. The results for $R = 0.5$ seem to represent a good overall compromise.

These results appear to show the BJT as a very promising device for high efficiency linear RFPA applications. The ability to use a circuit element to perform the linearization function on the transfer characteristic, as discussed in Section 1.3, is an asset which would not be available for a true transconductive device such as an FET. There is an important caveat on the above analysis, however. The junction has been assumed to be entirely resistive, albeit highly nonlinear. In practice the junction of an RF BJT will be shunted by a capacitance. Depending on the relationship between the frequency of use and the maximum frequency of the device, the impact of the junction capacitance on this analysis could be substantial. The situation is analogous to the discussion presented in *RFPA* (Chapter 5) concerning the effect of output capacitance of an RF power device in Class AB applications. In modern wireless communications, it is not uncommon to use a much higher frequency technology for applications below 1 GHz. A designer of HBT handset PAs using a 30-GHz HBT or pseudomorphic high electron mobility transistor (PHEMT) process could well find that the junction capacitance tends towards the low-impact extreme; a 2-GHz high power Si BJT, on the other hand, will almost certainly present a base-emitter impedance that is difficult to distinguish from a capacitor.

It is necessary, therefore, to repeat the above analysis for the opposite extreme case, where the device junction capacitance is sufficiently large that it can be assumed to act as a bypass capacitor for all harmonic current components, and is resonated out at the fundamental by the input matching network. The modified schematic diagram is shown in Figure 1.21. It is assumed now that the input-matching network effectively places a shunt resonator across the junction, such that the voltage across the junction is forced always to be sinusoidal. The input-matching network will also transform the generator impedance from its nominal 50-Ω value, down to form the input resistance R; R may also include some parasitic on-chip resistance. This circuit can be analyzed more easily. The input sinusoidal generator amplitude V_{in} will cause a corresponding change in the sinusoidal amplitude V_s appearing across the junction. Thus, if V_s is used as the chosen input signal

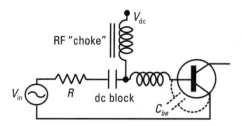

Figure 1.21 Schematic of BJT Class AB RFPA; junction capacitor shorts harmonic current components, but is resonated at fundamental by input matching network.

amplitude variable, it is possible to determine the current flowing in the junction,

$$ i_b = \frac{e^{k\left(V_q + V_s \sin at\right)}}{e^k} $$

where v_q is the dc voltage bias.

This can be integrated over a cycle in order to extract the fundamental component, I_{b1}.

The generator voltage amplitude required, then, to satisfy the defined condition is

$$ V_{in} = RI_{b1} + V_s $$

Figure 1.22 shows a comparable plot to Figure 1.18, showing a significantly modified set of waveforms for the same quiescent bias settings. The power sweep plot in Figure 1.23 can also be directly compared to the ideal junction plots shown in Figure 1.20. For a resistance setting of $R = 0.5$, it is clear that a higher quiescent current is required in order to achieve comparable linearity in the "harmonic short junction" case. This results in lower efficiency. However, a higher value of $R = 1$, shown in Figure 1.24, offers a somewhat better tradeoff, showing a peak efficiency just under 70% for comparable linearity and power gain. It is also significant that the more peaked waveforms in this case result in about 1-dB lower fundamental power than the comparable ideal junction analysis. Although the high-capacitance device shows less than the stellar performance of the ideal device, a practical device could be expected to give results somewhere between these two extreme cases.

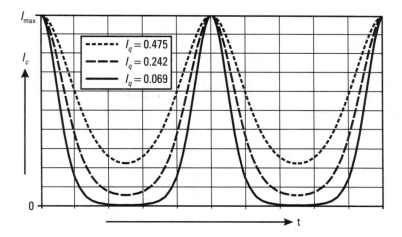

Figure 1.22 BJT "Class AB" current waveforms; junction harmonic short.

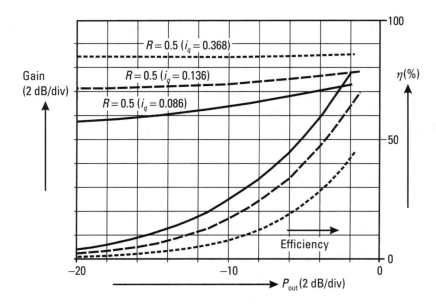

Figure 1.23 BJT Class AB gain and efficiency; junction capacitance assumed to act as harmonic short.

The analysis in this section has attempted, through the use of idealized but realistic device and circuit models, to shed some new light on the operation of an RF BJT power amplifier. Not only has it been shown that RF BJTs

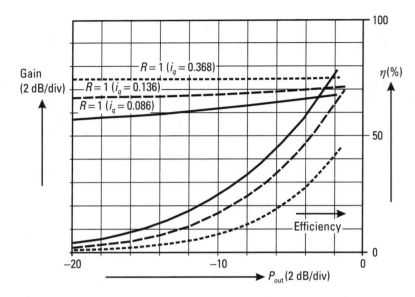

Figure 1.24 Gain and efficiency plots, harmonic shorted junction, higher ($R = 1$) series resistance value.

can make linear, highly efficient PAs, but they appear to have design flexibility which makes them arguably superior to the more ubiquitous FET devices. But in order to harness this potential, several basic design issues must be observed, namely:

- The input matching configuration, including the bias circuit, has a major impact on the operation of a BJT RFPA.

- The input match will show different optima for maximum gain, best linearity, and highest efficiency. Optimization of the second two may involve substantial reduction in power gain.

- Correct handling of harmonics is a necessary feature on the input, as well as the output, match. Situations where a device is being used well below its cutoff frequency may require, or will greatly benefit from, specific harmonic terminating circuit elements on the input.

- The process of linearizing the response of a BJT includes the use of a specific, and very low, impedance for the base bias supply voltage. This is a very different bias design issue in comparison to the simple current bias used in small signal BJT amplifiers, or the simple high impedance voltage bias used in FET PAs.

- The use of on-chip resistors in order to improve the linearity of a BJT RFPA device, as opposed to thermal ballasting, seems worthy of more extensive simulation and development efforts.

1.5 On Sweet Spots and IM Glitches

Sections 1.3 and 1.4 have presented a somewhat radical approach to the design of Class AB RFPAs. Essentially, the possibility has been demonstrated to "prescribe" an ideal device characteristic which will display the efficiency advantages of conventional Class AB modes, but have greatly suppressed odd-degree distortion. In Section 1.4, it was shown that some interesting possibilities exist for implementing this approach through the use of external circuit elements. None of this, however, is of much immediate help to a designer using the FET device technologies which dominate PA design above 1 GHz. Device technologists will not typically be able to respond quickly to requests for draconian changes in their device characteristics.

On the other hand, it is a matter of common experience that some FET device types show helpful "glitches" or "suckouts" in their intermodulation (IM) or adjacent channel power (ACP) responses. The origin of this behavior can be traced along much the same lines that were followed in Section 1.3; nonlinearities in the transfer characteristic can fortuitously cancel the non-linearities which are fundamental to Class AB operation. In many practical cases, the quiescent current setting for a particular device will be largely determined such that an IM "notch" is placed strategically near the maximum peak envelope power (PEP) drive level, so that the efficiency specification can be met.

Figure 1.25 shows a simple example of a sharp turn-on FET characteristic which has an additional gain expansion component in its "linear" region. The transfer characteristics of this device are

$$i_d = I_{max}\left\{ g_z v_{in} + \left(1 - g_z\right)v_{in}^{\,2} \right\}$$

showing a simple third-degree gain expansion term. The expression is normalized through the parameter g_z, which determines the amount of gain expansion but maintains the current such that at $v_{in} = 1$, $i_d = I_{max}$.

Such a device, operating at an appropriately selected Class AB quiescent bias setting, will show a sharp null in its IM characteristics, as shown in Figure 1.26. Note that even outside of the null, the IM3 is still substantially reduced in comparison to the linear ($g_z = 1$) case. Essentially, the gain

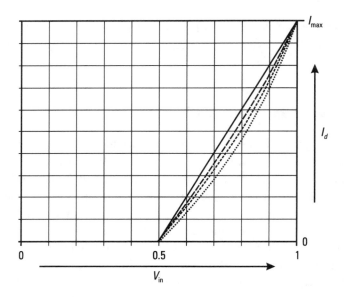

Figure 1.25 FET characteristic with gain expansion g_z parameter (see text) 1, 0.85, 0.75, 0.65.

compression caused by the truncation of the current waveforms in Class AB operation will be cancelled by the gain expansion built into the device characteristic. A simple third-degree expander such as this can only cancel or reduce third-degree effects. In practice, any device having gain expansion will have higher-degree components, and multiple nulls in higher order IMs are often observed. Figure 1.26 also shows a couple of apparent downsides; the power gain is reduced, surprisingly, as the gain expansion factor is increased. This is due to the reduction in gain at low drive levels and is actually more an artifact of the normalization of I_{max} than being a fundamental tradeoff. Figure 1.26 also shows the ideal linear device as having no IM or distortion at drive levels lower than the onset of Class AB truncation. This simply is a result of using an ideal linear transconductance ($g_z = 1$).

It is important to recognize that these effects may not in practice be simply attributed to the nonlinearity of the transfer characteristics. Nonlinearity in the device parasitics, especially the junction capacitances, can also play a part in creating sufficient nonlinearity in the device gain characteristics to provide cancellation of the Class AB compression at a specific power level. Such nulling phenomena can be difficult to preserve on a wafer-to-wafer basis, and over an extended period. But these effects demonstrate the general principle that device characteristics can be tailored to increase the linearity of efficient Class AB amplifiers.

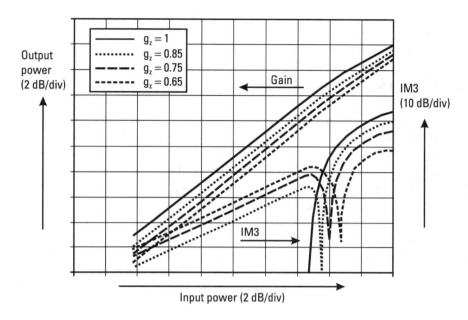

Figure 1.26 Gain and IM3 response for three values of g_z parameter.

IM3 notching will be discussed further in Chapter 3, and also in Chapter 5 in relation to the use of a predistortion device to reduce the IM levels. The gain expansion characteristic of a device could be viewed as a form of predistortion, which partially cancels the distortion caused by Class AB operation. From a more abstract mathematical viewpoint, the use of such a predistorter will inevitably generate higher odd-degree distortion terms, even though in the simple example described here the predistorter only has a third-degree coefficient. From the mathematical viewpoint, the nulling of the third-order IM products can be represented analytically as the cancellation of IM3 components arising from the third- and fifth-degree nonlinearities in the composite nonlinear response. The explanation given here is in effect a lower level of abstraction, where the physical origins of the different nonlinear processes are considered.

1.6 Conclusions

This chapter has attempted to show that there is more to Class AB amplification than may at first appear. The classical description assumes an ideally linear device, and then performs some crude waveform surgery in an attempt to harness the value of even degree distortion. A more logical approach in the modern era, where semiconductor wafers can be prescribed in three dimensions with close to molecular precision, is to synthesize device characteristics which retain the advantages of the traditional approach but reduce the disadvantages. Conventional Class AB operation incurs odd degree nonlinearities in the process of improving efficiency. Mathematical reasoning shows that it is feasible to specify a device characteristic that increases efficiency all the way up to 78% by the use of only even order nonlinearities. Such a device will not generate undesirable close-to-carrier intermodulation distortion.

The bipolar device emerges from this analysis very favorably, so long as its parasitics can be minimized at the frequency of intended use. The pragmatist may well find much to question in the analyses and device models used in this chapter. Parasitic reactances, especially nonlinear ones, will affect the waveforms and the conclusions. But as PA device technology advances, it is relevant and appropriate to idealize device behavior. At lower frequencies in the RF spectrum devices having cutoff frequencies in the 30–50-GHz region can show performance quite close to ideal.

Reference

[1]   Cripps, S. C., *RF Power Amplifiers for Wireless Communications,* Norwood, MA: Artech House, 1999.

2

Doherty and Chireix

2.1 Introduction

In any discussions about RF power amplifier techniques for modern applications, the central goal of maintaining efficiency over a wide signal dynamic range must surely remain paramount. Yet the intellectual and technical challenges of understanding and implementing linearization methods seem to have stolen the limelight in recent years. Some, if not all, of the linearization goals which challenge the modern RF designer become relatively trivial if efficiency is removed from the equation; backed-off Class A amplifiers still take a lot of beating when linearity is the sole criterion. It is therefore surprising that several PA design techniques which date from a much earlier era, and which have demonstrably addressed the efficiency management issue, have been largely ignored by the modern RF design community.

Probably the least extreme case, in terms of neglect, has been the "Envelope Elimination and Restoration" (EER) method; this is widely attributed to Kahn, who published a paper on the technique [1] in the single sideband (SSB) era of the early 1950s. In fact, the application of high-level amplitude modulation (AM) to a Class C RFPA was common practice in the tube era, as any reference to contemporary ham radio literature will confirm. Kahn's innovation was essentially the generation of a constant-amplitude, phase-modulated signal component which could be amplified using a nonlinear PA. Coupled with the modern power of digital signal processing (DSP), EER provides one important avenue for the taming of steep downward efficiency/dynamic range curves exhibited by any Class AB amplifier.

The technique has an Achilles' heel—the need to convert a suitably profiled, linearized PA supply drive signal to the necessary high level of current and voltage required by the PA itself. This not only erodes the efficiency advantage, through the additional efficiency factor of a power converter, but also has an important impact on the maximum viable signal bandwidth. There is also a secondary issue of dynamic range; most RFPAs of standard design will display sufficient transmission, even with zero supply, to limit the dynamic range of the system to about 20 dB.

Successful implementations of EER which demonstrate some alleviation to these problems have appeared in recent literature [2]. The focus in this chapter, however, is to follow some alternative avenues in pursuit of a PA design technique which conserves efficiency in wide dynamic signal range applications; the methods proposed in two classical papers from the 1930s, by Doherty [3] and Chireix [4]. The Doherty technique emerges very favorably from this closer scrutiny. It seems to offer more than its protagonists have proposed to date, and even has claims as a linearization method in its own right. The Chireix outphasing method, on the other hand, does not seem to stand up quite so well to modern CAD analysis, but some of its elements are well worth studying.

2.2 The Doherty PA

2.2.1 Introduction and Formulation

The "classical" Doherty PA (DPA) was analyzed in some detail in *RFPA* (Chapter 8). The configuration analyzed here, Figure 2.1, still uses two active devices but assumes a more generalized transistor transfer characteristic. So the basic elements are the two devices themselves, an impedance inverter, and a common RF load resistor. The impedance inverter can be considered conceptually to be a simple quarter-wave transmission line transformer, whose terminal characteristics have the form

$$\begin{bmatrix} V_2 \\ I_2 \end{bmatrix} = \begin{bmatrix} 0 & jZ_o \\ 1/jZ_o & 0 \end{bmatrix} \begin{bmatrix} V_1 \\ I_1 \end{bmatrix} \tag{2.1}$$

although a practical implementation may beneficially use other networks to achieve this functionality. The active devices are, for the purposes of this analysis, assumed to be conducting different fundamental current amplitudes, I_m and I_p, at any given input signal amplitude v_{in}, where

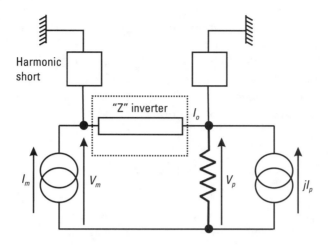

Figure 2.1 Schematic for two-device Doherty PA.

$$I_m = f_m(v_{in}), \quad I_p = f_p(v_{in})$$

which are not necessarily simple linear functions of the input drive signal v_{in}; this is a substantial generalization of the simple case considered in *RFPA*, where ideal linear transconductive dependencies were assumed. Ideal harmonic shorts are assumed to be placed across each device, so that only fundamental voltage and current components are considered in the analysis.

Referring to the nomenclature of Figure 2.1, (2.1) can be expanded to give

$$V_p = jZ_o I_m \qquad (2.2)$$

$$I_o = \left(\frac{1}{jZ_o}\right) V_m \qquad (2.3)$$

and the remaining circuit relation is

$$I_o = jI_p - \frac{V_p}{R} \qquad (2.4)$$

We require expressions for the voltages at each device output, V_m and V_p, in terms of the device currents I_m and I_p; clearly (2.2) gives one such relationship straight away and shows that the peaking device output voltage, which is one and the same as the final output load voltage, is proportional to the main device current, I_m; in other words it is *independent of the value of I_p*. Thus, provided the main device voltage is kept below clipping levels by the action of the peaking device, the linearity of the final assembly is defined entirely by the main device characteristic $f_m(v_{in})$. The requirement for the peaking device to perform this function can be determined from (2.3) and (2.4),

$$V_m = Z_o \left[\left(\frac{Z_o}{R} \right) I_m - I_p \right] \tag{2.5}$$

showing the possibility of using a suitable peaking device characteristic $f_p(v_{in})$ to "neutralize" the rapidly rising main device voltage and keep it below clipping level over the entire input signal dynamic range.

Given that this neutralization process is clearly feasible through (2.5), (2.2) emerges as a remarkable result; its significance can easily be missed in the typical idealizations used in the analysis of a standard Doherty configuration. The key point is that whatever gyrations the $f_p(v_{in})$ function makes in order to perform its task, and however approximate, imperfect, or nonlinear they may be, the action of the peaking device remains invisible at the output load, whose voltage remains proportional to I_m.

This remarkable property verges on being a linearization process; in a typical configuration, most of the RF output power in the upper range is being supplied by the peaking device, which will most probably be a substantially nonlinear device. The dependency of the output power on the input drive signal, however, remains as defined by the main device characteristic, which can be much more linear.

This property will now be further illustrated in some specific examples. Some further nomenclature needs to be defined, in order to allow for more generalized cases. Of particular importance in any Doherty PA will be the relative I_{max} values for the main and peaking devices; this forms one of the essential starting parameters for a design. Throughout this analysis, the symbols I_m, I_p are the amplitudes of the fundamental components of the main and peaking device currents, and not their instantaneous values. This is common enough practice in ac analysis and does not need further justification, but it does pose some hazards for the unwary; this applies especially to the definition of "I_{max}" values for both devices, which will again refer to the maximum values of the fundamental component of current for each device, and not to its physical "I_{max}" value. For this reason, the "I_{max}" symbolism will be avoided:

$$\left(I_m\right)_{max} = I_M$$

$$\left(I_p\right)_{max} = I_P$$

$$\frac{I_P}{I_M} = \Gamma$$

The signal drive range will, as usual, be normalized so that

$$0 < v_{in} < 1$$

and the "breakpoint" at which peaking current starts to flow is defined to be $v_{in} = v_{bk}$, where clearly $0 < v_{bk} < 1$.

One of the features which will be explored in this section is the possibility of using devices which have a nonlinear characteristic, rather than the "linear-dogleg" assumption which is typically used. In this respect we are taking up a similar change in mindset as was applied and discussed at some length for Class AB modes in Chapter 1 (Section 1.3). In such cases, the breakpoint is effectively replaced by the nonlinear characteristics of the device and ceases to be a major design parameter.

2.2.2 The Classical Doherty Configuration

For a "classical" Doherty configuration,

$$I_P = I_M, \ (\Gamma = 1)$$
$$I_p = f_p\left(v_{in}\right) = v_{in}I_p \qquad 0.5 < v_{in} < 1,$$
$$\qquad\qquad = 0 \qquad\qquad 0 < v_{in} < 0.5$$
$$I_m = f_m\left(v_{in}\right) = v_{in}I_M \qquad 0 < v_{in} < 1,$$
$$v_{bk} = 0.5$$

With these ideal device characteristics, the choice of the circuit parameters R and Z_o is straightforward. The peaking device is inactive up to the breakpoint, and the R and Z_o values are chosen such that the main device voltage reaches its prescribed maximum allowable value at the breakpoint drive level, $v_{in} = v_{bk}$. This voltage maximum will usually correspond to the clipping level, which in turn is approximated by the dc supply; as with any PA design, some allowance will usually be included on the dc supply itself due to the knee or turn-on region of the device. There may be other considerations, given that the DPA action enables this level to be a design parameter, rather than a physical one. For clarity, however, we will assign the symbol V_{dc} to this voltage limit. Thus, at the breakpoint drive level, (2.5) gives

$$V_m = V_{dc} = (0.5)I_M \left(\frac{Z_o^2}{R} \right) \tag{2.6}$$

A second relationship between the circuit parameters can be obtained by considering the maximum drive condition, $v_{in} = 1$; from (2.5), the voltage amplitude at the main device is given by

$$V_m = Z_o \left[\left(\frac{Z_o}{R} \right) I_m - I_p \right]$$

the value of Z_o is therefore selected such that V_m has the same value at the maximum drive as it did at the breakpoint; so

$$V_{dc} = Z_o \left[\left(\frac{Z_o}{R} \right) I_M - I_P \right] \tag{2.7}$$

Equations (2.6) and (2.7) can be solved to give

$$R = \frac{V_{dc}}{2I_M}, \quad Z_o = \frac{V_{dc}}{I_M} \quad (= 2R)$$

noting that the load presented to the main device at the breakpoint is

$$\frac{2V_{dc}}{I_M} = 4R$$

Note that this value is a factor of two higher than would normally be used for a conventional linear design with no peaking device. This is a simple illustration of a novel way of viewing the action of a Doherty PA; the action of the peaking device is to enable the use of a much higher value of load resistor for a given main device yet still maintain similar transconductance linearity over the whole input drive range. The specific values for R and Z_o are predicated in this case through the idealized device characteristics. In practice, both values will be subject to some optimization. The main objective in this treatment of the subject is to show that useful Doherty action can be obtained for quite a wide range of real device characteristics, some of which do not even closely approach the ideal models chosen in this simplest and classical analysis.

Figure 2.2 shows the results of the above analysis, in terms of the voltages and currents at each device. The key feature is the maintenance of a constant voltage swing at the main device in the upper regime. This will result in high efficiency for the main device over the entire extent of the upper regime, and a much better efficiency/backoff characteristic than would be obtained from a conventional design. Efficiencies are not plotted at this juncture; until some more specific details are established about the implementation of the two amplifiers, it is not strictly possible to derive an efficiency curve. It is another focus of the present analysis to explore a wider range of possibilities, rather than just following the conventional assumption that the main device is an ideal Class B PA.

Any attempt to realize such a configuration runs into an immediate problem with the defined characteristics of the peaking PA. What appears to be required is a device which remains completely shut off up to the

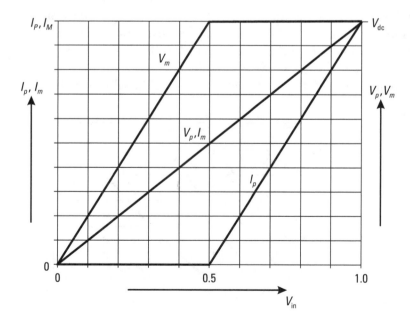

Figure 2.2 Classical Doherty PA; current and voltage (fundamental amplitudes) characteristics for main and peaking devices plotted against input drive signal amplitude.

breakpoint, which in this case is 6 dB backed-off from the maximum drive level. Not only that, but once conduction starts, it has to make up for lost territory and race up to the I_M value of the main device, with only half as much change in drive voltage. Radio frequency integrated circuit (RFIC) designers have some advantages in that they can consider a different doping or implant schedule for the peaking device. But even then, the problem of holding the device off over all but the upper few decibels of drive power remains. The most obvious and direct route to a possible solution is to use a device biased well beyond its cutoff—a Class C amplifier. There are several practical problems involved in doing this; the extent to which the device has to remain cut off requires "deep" Class C bias, which then raises breakdown issues. It also raises gain and device periphery issues. However, it is a starting point and was the original method used by Doherty, with tubes, to demonstrate this configuration.

In order to incorporate PAs with a wide range of cutoff and conduction angles, it is necessary to recall some of the results on the basic analysis of these modes. Figure 2.3 shows a plot of the fundamental component of current versus sinusoidal drive voltage for various quiescent bias settings for a device having the conventional ideal "linear-dogleg" characteristic defined in Chapter 1. At first sight, the Class C ($V_q < 0$) curves seem to show useful possibilities for realizing the peaking device characteristics, assuming that the main device followed something close to the Class B ($V_q = 0$) curve. A quiescent bias, for example, of $V_q = -0.5$ would result in something very close to the required "holdoff" range shown in Figure 2.2. But if both devices are

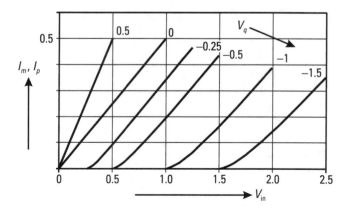

Figure 2.3 Fundamental current component for reduced conduction angle mode device; device has linear conduction range $0 < V_{in} < 1$; quiescent bias settings (V_q) measured in same units as V_{in}.

identical and have the same input level, the Class C peaking device would only reach about 40% of the required magnitude at the maximum drive point, $V_{in} = 1$. So to make this approach work, the periphery (i.e., the I_P value) of the Class C device must be increased, by a factor of about 2.5, even if ideal transconductive characteristics are assumed.

It turns out that the peaking current drive problem is less severe when a more generalized range of DPA cases is considered; it is possible to compromise the requirements of peaking current without losing the efficiency advantages of the configuration too quickly. But realization of the peaking PA characteristic is a problem which appears to get worse as more realistic device models are considered. This issue probably represents the main practical stumbling block for implementation of Doherty PAs in the modern era. There has been a justifiable reluctance to tackle this problem using "smart" bias adaptation, or switch/attenuator controls, because such techniques instantly corrupt the elegant simplicity of a "self-managing" DPA. It appears, however, that this is a step which must reluctantly be taken.

This will be pursued in due course, but for the present it is instructive to complete the analysis of efficiency assuming the use of an ideal Class B ($V_q = 0$) main device and a Class C ($V_q = -0.5$) peaking device with appropriate I_p scaling. The voltage, current and efficiency curves are shown in Figure 2.4. The details of the analysis have been omitted; it is essentially a

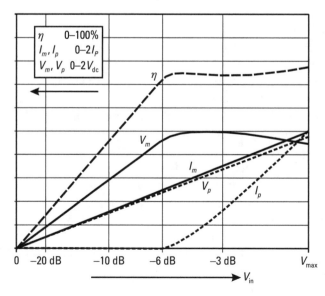

Figure 2.4 RF current, voltage amplitudes and efficiency for Doherty PA using Class C peaking device.

matter of extending the reduced conduction angle analysis shown in Figure 2.3 to include the dc components. The result is essentially the classical Doherty efficiency curve for a symmetrical configuration and a 6-dB breakpoint. The slightly sluggish start to the peaking PA conduction, caused by the Class C operation, causes some variation in the main PA voltage beyond the breakpoint. This also necessitates reducing the Z_o value slightly in order to keep the main PA voltage swing below the stipulated supply rail value. The key feature is the return to Class B efficiency at the breakpoint, where the peaking device shuts down; thereafter as the drive power is further reduced, the efficiency drops down the conventional curve for a Class B PA.

2.2.3 Variations on the Classical Configuration

As an introduction to more generalized, and practical, implementations of the Doherty PA, Figure 2.5 shows the use of a nonlinear device as the peaking PA. This is picking up a theme from Chapter 1 (see Section 1.3) where the concept of using a nonlinear characteristic, rather than using the cutoff of a linear device, was explored for Class AB applications. In order to illustrate an important point, a somewhat arbitrary device characteristic is shown. The two device characteristics are plotted in Figure 2.5; clearly, there is a distinct "violation" of the original requirements for the peaking device, in that the peaking device never fully shuts down, and is palpably nonlinear. The voltage curves, therefore, tell an interesting story. First and foremost, the peaking device voltage V_p still shows the ideal linear response pertinent to Class B operation. The main device voltage shows only an approximation to the required characteristic. The less abrupt main device voltage "cornering" will have some impact on the efficiency, as shown in Figure 2.5. As always in DPA analysis, the efficiency curve does not follow implicitly from the voltage and current relationships and depends on the more detailed implementation of the peaking device characteristic; in this case an ideal cube-law device at zero quiescent bias has been assumed for the peaking device.

Figure 2.6 shows the efficiency backoff plot for three cases: ideal Doherty using a Class C peaker with suitable periphery and bias offset adjustment, a simple Class B PA, and the less-ideal peaker realization shown in Figure 2.5. It will be seen that the efficiency impact for the nonlinear peaker is significant but not by any means fatal. But the other issue is to recognize that the nonlinearity of the peaking PA does not show up in the final transfer characteristic; the function of the peaker is to pull the voltage at the main PA lower than the maximum value stipulated by the rail supply. Indeed, once the basic principle that linearity is maintained regardless of the

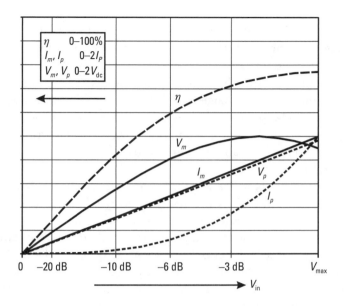

Figure 2.5 RF current, voltage amplitudes, and efficiency for Doherty PA using nonlinear peaking device.

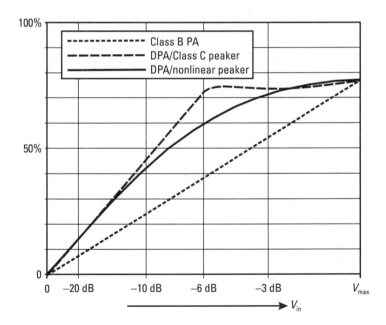

Figure 2.6 Efficiency characteristics: ideal Doherty, Doherty using nonlinear peaking device (see text and Figure 2.5); and conventional Class B performance shown for comparison.

peaking current characteristic, one can quickly drift into a mode of thinking which verges on the cavalier; almost anything will do, within some rather broad boundary conditions. It raises the important concept that there is really no requirement that the main and peaking devices need to be matched in any way; different technologies, FET/BJT mixes, simple analog bias and gain adaption, all become possibilities for realizing the peaking PA function. Before pursuing some of these possibilities, however, it is worthwhile to consider an important generalization of the basic configuration, that of the asymmetrical Doherty PA.

The key point about the so-called asymmetrical DPA is that the breakpoint can be stipulated. This will result in devices having nonequal values of I_P and I_M, and is highly relevant in applications where the PEP is much higher than the mean power such as in most modern communications systems. Figure 2.7 shows the current and voltage backoff relationships for a typical asymmetrical DPA. In this case, the breakpoint is a much lower backoff point than the classical 6-dB breakpoint. It can be seen that to achieve this, the two devices have different maximum currents, so that the ratio between I_P and I_M, defined above as Γ, now becomes an additional design parameter. The determination of design values for R, Z_o, and Γ is, however, still quite straightforward. The main device voltage, V_m, at the breakpoint will be defined to reach the stipulated clipping level, V_{dc},

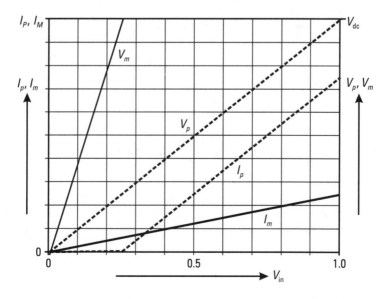

Figure 2.7 Asymmetrical Doherty PA; current and voltage characteristics for main and peaking devices plotted against input drive signal amplitude.

$$\left(V_m\right)_{v_{bk}} = V_{dc} = v_{bk} I_M \left(\frac{Z_o^2}{R}\right) \tag{2.8}$$

and also at the breakpoint the peaking device voltage V_p will have climbed only to a magnitude

$$\left(V_p\right)_{v_{bk}} = v_{bk} V_{dc} \tag{2.9}$$
$$= v_{bk} I_M Z_o$$

from (2.2).

Equations (2.8) and (2.9) give values for R and Z_o,

$$R = v_{bk}\left(\frac{V_{dc}}{I_M}\right), \quad Z_o = \left(\frac{1}{v_{bk}}\right)R \tag{2.10}$$

The Γ value can be determined by equating the main device voltage at the breakpoint and the maximum drive point, from (2.2) and (2.5),

$$Z_o\left[\left(\frac{Z_o}{R}\right)I_M - I_p\right] = v_{bk} I_M \left(\frac{Z_o^2}{R}\right)$$

whence

$$\Gamma = \frac{I_P}{I_M} = \frac{1 - v_{bk}}{v_{bk}} \tag{2.11}$$

Figure 2.7 shows the case for $v_{bk} = 0.25$, representing a breakpoint at 12-dB backoff. This will require a Γ ratio of 3. In a modulation system which has occasional peaks rising up to 12 dB above the average power level, such a configuration would only utilize the much larger periphery peaking device when required to transmit a peak, thus saving much power. Once again, it is not possible to plot an efficiency curve without making some stipulations about how the peaking amplifier will be realized. Referring to the Class C transfer curves in Figure 2.3 ($V_q < 0$ cases), it is clear that the peaking device periphery must be scaled up, by at least the Γ factor. On the other hand, the requirement for a lower breakpoint value does somewhat alleviate the periphery requirements of the peaking device if a simple Class C approach is used.

Figure 2.8 shows such a realization, which now allows the efficiency to be computed. Compared to the ideal characteristics shown in Figure 2.7, some adjustment has to be made to the Z_o value in order to keep the main device voltage below the allowable maximum value. The efficiency curve still looks quite attractive, despite an inevitably deeper dip in the mid-backoff region. Such an asymmetrical DPA would, on most counts, appear to be a more worthwhile target than the classical symmetrical version, for practical realization in modern applications.

The asymmetrical DPA would normally be associated with high peak-to-average power ratio signals, such as occur in wideband CDMA (WCDMA) or multicarrier systems. In such applications, the use of a peaking device with several, or many, times the peak current of the main device can greatly improve overall efficiency. There is, however, another side to the asymmetrical coin; there may be some benefit in having a peaking device with a lower peak current. One of the reasons for examining this concept is that in practice this will be the case if two similar devices are used, in an attempt to realize a classical configuration. Looking again at Figure 2.3, if the main device is run in Class B ($V_q = 0$), and a similar device is biased into Class C (say, at $V_q = -0.5$), it is clear that at the drive level required to obtain

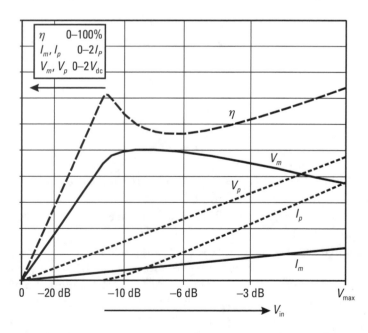

Figure 2.8 Asymmetrical Doherty PA; peaking function realized using Class C and scaled periphery device.

the maximum peak current in the main device, the peaking device is only producing a fundamental component of about 40% of the main device component, I_M. The result of this is that the peaking device current will always be insufficient to pull down the main device voltage, and therefore a lower load resistance value must be presented to it, by reducing the Z_o value. This situation is shown in Figure 2.9, and is here termed *Doherty-Lite*.

The key point about the Doherty-Lite configuration is that a significant improvement in backoff efficiency can be obtained with the simplest possible circuit configuration. The efficiency plot will not, however, have the classical "twin peaks." We will see in Section 2.2.4 that there is a reality to be faced with any of the "heavier" Doherty PAs described thus far, and in other literature. Practical realization of the peaking PA function by scaling of the device periphery and Class C biasing runs into an escalation of practical difficulties. Alternatives, in the form of bias adaption and DSP amplitude control, need to be pursued. But the Doherty-Lite approach gives some more modest benefits whilst retaining a viable and simple practical circuit. Figure 2.10 shows the peaking current and efficiency characteristics for three bias settings of the peaking device. The benefits of Doherty-Lite operation seem

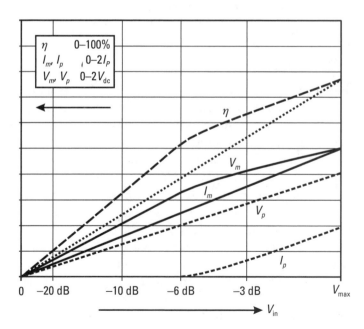

Figure 2.9 "Doherty-Lite" PA using same device type for main and peaking functions. (Unlabeled dotted line is conventional Class B efficiency, for comparison. Normalized circuit element values: $R = 1$, $Z_0 = 1.625$.)

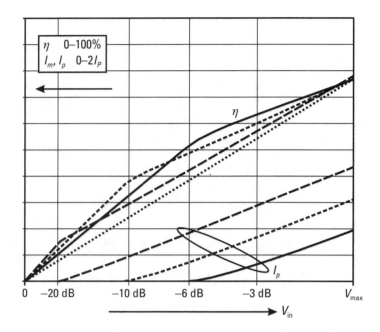

Figure 2.10 "Doherty-Lite" PA; various bias settings for peaking device: $V_{pq} = -0.5, -0.3,$ and -0.1. Z_o is adjusted in each case. Dotted line shows conventional Class B efficiency characteristic.

to have a broad optimum in the vicinity of $v_{pq} = -0.3$; this gives less efficiency improvement in the upper drive range, but quite substantial improvements in the -10-dB backoff region, in comparison to a conventional Class B arrangement. The power contribution of the peaking device will also be greater as the breakpoint is set to lower PBO values.

At this point, we seem to have a configuration which is sufficiently viable in a practical sense, to run a full scale simulation using "real" device models. Figure 2.11 shows such a simulation, where the main and peaking devices are identical, but the main device has been optimized for a reasonable efficiency/linearity compromise. The peaking device is biased so that it becomes active at about the -10-dB backoff point. There is clearly a major efficiency improvement in the Doherty-Lite case, amounting to about 20% in the midrange, and 10% over most of the 20-dB range. The peaking device adds about 2 dB, rather than the full 3 dB which would be expected from two equal periphery devices. There is also a small linearity degradation evident on the Doherty plot. This can be traded with efficiency by fine-tuning the RF load and Z_o values. This simulation, which includes detailed models

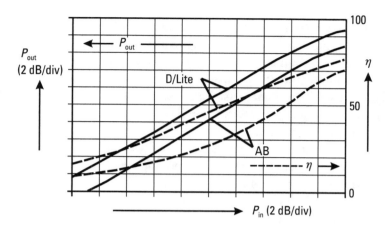

Figure 2.11 Simulation of Doherty-Lite PA using two identical GaAs MESFET devices. For comparison, the performance of the main device by itself is also shown, at the same bias.

for the devices including knee effect and output conductance, is a reassuring result; the Doherty principle passes a modern simulator test quite well, and the challenge remains to devise suitable peaking amplifier configurations.

2.2.4 Peaking Amplifier Configurations

It has been established, over the last few sections, that the major stumbling block in realizing the full benefits of a DPA is the peaking amplifier function. Indeed, it can be speculated that this is really the core reason that the DPA has been so slow to come into more general use in the modern multicarrier power amplifier (MCPA) era. The problem has been greatly multiplied by a reluctance to concede that the simple approach of using input bias offset causes too many problems, and has to be abandoned in favor of more complex adaptive schemes. This reluctance is well founded. Just as with linearization systems, the intrusion of video processing greatly reduces the potential speed of a technique, and may even restrict the use to system level, rather than RF level, implementation.

Clearly, at the system level, a transmitter "knows" everything about the signal which it is about to send to the RFPA. It can therefore be fairly assumed that it will not be much of an extra burden for the system processor to generate, for example, a voltage which is proportional to the RF envelope amplitude. Such a signal can be put to immediate use in a bias adaption scheme and the peaking amplifier problem almost solves itself. A standalone

PA, on the other hand, can only obtain such information using some form of envelope detector. Such a detector may require tens of RF cycles in order to measure their peak value; the number of cycles required is in turn a function of the precision required. Fortunately, unlike in linearization schemes which will be discussed in later chapters, the precision requirements for an envelope detector in this application are very forgiving.

Figure 2.12 shows the very modest requirements of a bias adaption scheme for a peaking PA. The device is biased beyond its cutoff point, as in a conventional fixed bias configuration. Once the RF drive reaches the peaking PA breakpoint, however, the bias point is shifted monotonically with the increasing RF drive level, so that once the drive level has reached the maximum level, the peaking PA is biased at the same point as the main PA. The key point about this scheme is that the two devices can have the same periphery, assuming a classical symmetrical DPA with a 6-dB breakpoint. In an asymmetrical case, the peaking device does not need the large extra periphery factor that would be needed with fixed bias. Figure 2.12 also shows that the

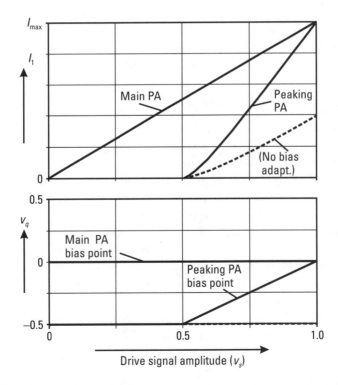

Figure 2.12 Bias adaption scheme for peaking PA realization in a classical Doherty configuration.

linearity of such a device is actually no worse than it would be in a Class C configuration; in any case, it has already been determined that the linearity of the peaking device does not play a primary role in determining the overall linearity of the combination. So although Figure 2.12 shows a linear bias shift with input envelope amplitude, some nonlinearity could be tolerated in this characteristic. This would be important if the bias adaption was being controlled by a detector.

Figure 2.13 shows a simulation of a classical Doherty PA using this form of bias adaptation; the two devices in this simulation are now identical types. One additional feature of having a bias adaption scheme of this kind is that the mode of operation of both amplifiers is much more flexible, and a much wider range of possibilities opens up. For example, the simulation results in Figure 2.13 show a main PA operating in deep Class AB, and the peaking PA adapting to the same bias setting at maximum drive level. Figure 2.14 returns to the more idealized analysis and shows some other possible variations, where the main PA is operated in Class A. The peaking PA starts off in Class B and ends up in Class A due to the bias adaption. Although the efficiency plots are not as "stellar" as typical DPA plots based on Class B operation, the efficiency in the backed-off region is anywhere between two to five times the efficiency of a conventional Class A amplifier, with reasonable expectation of comparable linearity.

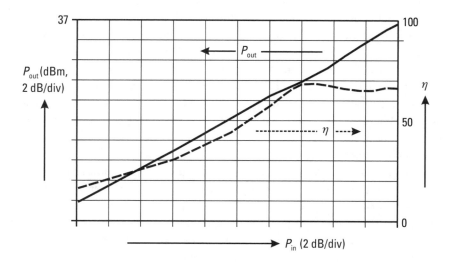

Figure 2.13 Simulation of Doherty PA using two identical GaAs MESFET devices; bias adaption used to obtain equal I_{max} values.

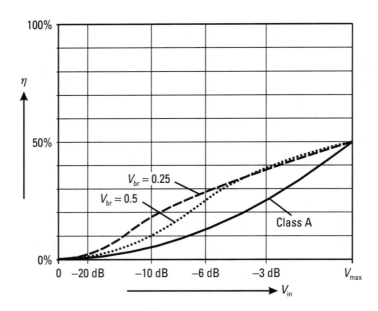

Figure 2.14 Efficiency curves for Doherty PAs using Class A main PA; conventional PA shown for comparison.

2.2.5 Doherty PA Matching Topologies

In the DPA circuits analyzed so far in this section, it has been assumed that the final RF load is a resistor of appropriate value. This is a convenient simplification for analysis, and in principle only omits the details of a simple matching network in the final practical realization. So long as the harmonic shorts at each device are forcing a sinusoidal voltage, this will be a valid simplification. In the case of the DPA, however, it is necessary to exercise considerable care in applying this final step towards the realization of a practical circuit. There is the additional issue that if quarter-wave SCSS harmonic traps are used, the third harmonic generation may cause some changes to the overall circuit performance.

Figure 2.15 shows one possible practical configuration. The main PA device output network is realized by using the impedance inverter to perform the additional function of matching, in a conventional quarter-wave transformer configuration. The peaking device poses a trap for the unwary, in that no inversion is required but most conventional matching networks will supply an inversion function. This can be overcome by the use of an additional quarter-wave section, as shown. This section can be absorbed into the matching function or, as shown, the matching can be provided entirely by the first

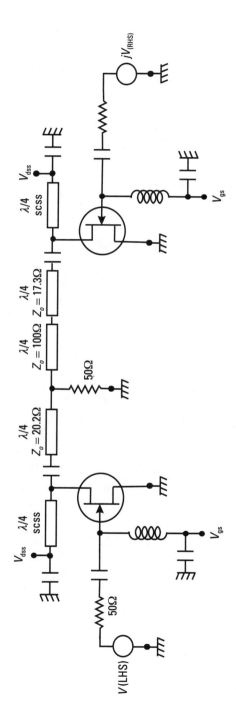

Figure 2.15 Schematic for practical DPA realization.

quarter-wave transformer. The impedance transformers in Figure 2.15 have been designed to give the same device impedance environment as that shown in the simulation of Figure 2.13. This used a final load resistance of 1.5Ω, so that in the backed-off regime where the peaking device is inactive, the main device load impedance is

$$\frac{3.5^2}{1.5} = 8.2\Omega$$

so that the required impedance transformer has a characteristic impedance of

$$\sqrt{50x8.2} = 20.2\Omega$$

On the peaking side, the impedance requirement can be most easily reckoned by noting that at the maximum drive point both devices are generating equal and in-phase fundamental components so that the load is evenly split between the two sides. The peaking side can therefore be designed as if it was independently working into a 100-Ω load. The required transformer characteristic impedance is therefore given by

$$\sqrt{100x3} = 17.3\Omega$$

and the phasing inverter will require a 100-Ω characteristic impedance.

The final output topology looks, therefore, at first sight to be "mirrored" from that usually shown in a classical DPA analysis due to the inverting action of the matching transformers. Figure 2.16 shows the simulated responses of the matched DPA, which can be seen to be very similar, but not identical, to those shown in Figure 2.13 for the more idealized broadband load version. Although in this example the necessary transformer characteristic impedances could be considered quite realizable, this will be less likely in higher power applications. In these cases the matching function for higher power devices will probably have the lowpass lumped element form shown in Figure 2.17. These networks also have an inverting action, especially when being used to transform high impedance ratios. This can be shown by recalling that the resistive part of the transformed impedance, R_T, shown in Figure 2.17 is given by

$$R_T = \frac{R_o}{1 + \left(\dfrac{R_o}{X_C}\right)^2}$$

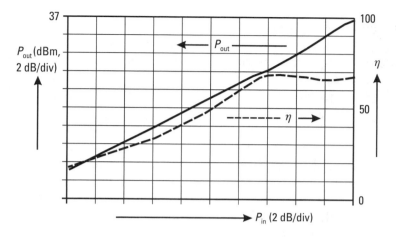

Figure 2.16 Simulation of "matched" version of Doherty PA.

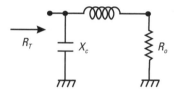

Figure 2.17 Lowpass matching network.

where X_C is the capacitive reactance of the shunt matching capacitor. So for larger transforming ratios, where

$$\left(\frac{R_o}{X_C}\right)^2 \gg 1$$

$$R_T \approx \frac{X_C{}^2}{R_o}$$

which has the same inverting and transforming function as a quarter-wave line, with the capacitive reactance replacing Z_o in the expression for impedance. In lower transformation ratio applications, a lowpass lumped element equivalent is still possible using a symmetrical "pi" section (see *RFPA*, p. 83).

Another practical requirement is for a 90° phase compensation between the two devices, to allow for the additional 90° phase shift which now appears on the peaking side. This can most conveniently be realized using a quadrature 3-dB hybrid to divide the inputs, although care is required to ensure that the correct phasing is employed. Unlike a conventional balanced amplifier, which uses quadrature couplers on input and output, the orientation of coupled and direct ports on the input connection is no longer arbitrary.

One final point should be made concerning the simulation shown in Figure 2.16. The model employed is that of a GaAs MESFET which has a particularly well-behaved characteristic in comparison to some other devices which might be considered for applications below 2 GHz. In particular, the MESFET has very low output capacitance, which has been assumed to be tuned out by the output matching network. This is a valid assumption provided that the output capacitance shows only small variations in between its "on" and "pinched-off" conditions. This may become a more troublesome issue with some device types, and at higher frequencies. Essentially, the peaking device in its pinched-off condition needs to look like a comparatively high impedance. In some cases, it may be necessary to design, or tune, the peaking device output match such that it provides a compromise between power and efficiency in the "on" condition and high impedance in the "off" condition.

2.2.6 The Multiple Doherty PA

The extensions and added flexibility offered by both asymmetrical DPA and bias adaption would seem to cover a large range of options and applications. It is therefore noted only in passing that yet another dimension can be considered, that of using more than two devices in a Doherty configuration. The principle is illustrated in Figure 2.18. The "main" device is now a DPA in itself. The analysis becomes increasingly cumbersome and is available elsewhere [5]; it seems that experimental results have not been published, and it would be a fair speculation that such amplifiers have never been built successfully in the microwave frequency range. In principle, the use of more devices enables a flatter efficiency PBO characteristic to be obtained over a wider range of PBO, but the practical issues of managing the bias offsets and the increasing circuit complexity are obvious detractions. Until such time that successful and useful implementation of the two-device DPA has been demonstrated, this would appear to be a step in the direction of diminishing returns.

2.2.7 Doherty PA Conclusions

The analysis and simulations in this section have given a clear indication that the Doherty technique is not only viable and potentially useful in modern

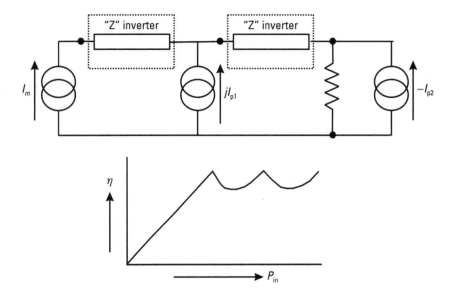

Figure 2.18 Schematic for multiple-device Doherty PA.

applications, but has some interesting variations that have not been well covered in the literature. It is important, however, at this concluding stage, to recognize that some issues remain unresolved. The tradeoff between linearity and efficiency seems to be very fundamental, and must be respected. The concept has been promoted in this section that the linearity of a DPA is entirely determined by the main PA and is independent of the linearity of the peaking PA. This is only true inasmuch as the linearities are assumed to be entirely transconductive. In practice, and as shown clearly on the simulated plots, the linearity of the main PA below the breakpoint will be degraded as a result of operating at higher voltage swing, for a given power level, than in normal operation. Nonlinear effects such as IM and ACP are caused by output conductance nonlinearities as well as transconductive effects; the conductance variations can be reactive as well as resistive. These considerations all suggest that some device types may behave much better than others in DPA mode. Although in practice the linearity of a DPA may not be as good as a conventional Class AB design using a comparable device, the analysis in this section has shown that this traditional tradeoff can possibly be outflanked; a device whose linearities are primarily transconductive can, in principle, be used to "linearize" a peaking device which delivers most of the peak power. As a minimum, there appear to be possibilities worthy of further research in this area.

2.3 The Chireix Outphasing PA

2.3.1 Introduction and Formulation

Section 2.2 has shown that at least one old PA technique can be revived to solve modern problems; this particular old dog seems able to learn some new tricks. Application of CAD tools not only confirms the viability of the original technique, but also shows some new and useful variations. Sadly, this is not the case with all such older methods. Some old dogs seem more reluctant, to the extent that one has to wonder whether their tricks of old were quite so good as contemporary reports suggest. This certainly seems to be the modern perception on the Chireix outphasing PA technique, first described in a much quoted, but famously indigestible paper [4].

Like the Khan EER approach, the Chireix method is not a PA as such; it is a transmitter architecture which constructs a high-level amplitude-modulated signal. It therefore assumes that the necessary information about the signal is available, either in analog or digital baseband form. It also has a central plank which is frequently mis-assigned. The construction of an amplitude modulated signal using a pair of cw sources having a variable differential phase offset is a technique used widely in ac systems and is not unique to the Chireix PA. The key additional component which Chireix claimed was the ability to reduce the dc power drawn by a pair of high-efficiency PA devices when in a condition of high "outphasing," that is to say, when the two outputs are being nearly phase-cancelled in order to generate a signal at the low end of the dynamic range. This process relies, like the Doherty PA, on the load-pulling effect of one device on the other. It is this aspect of the Chireix technique which seems to be least appreciated and forms the focus in this section. Although transmitter, as opposed to PA, techniques are somewhat ruthlessly excluded in this book and *RFPA*, this particular technique requires more in-depth understanding of the PA component itself and is therefore worthy of a more detailed review.

Figure 2.19 shows a basic configuration. Two devices are operating into a common, differentially connected load resistor. There is a variable phaseshifter, shown as a differential element on each device input; each element has at least a 90° range. Clearly, if the differential phase offset between the two device inputs is set to be 180°, conventional differential, or push-pull, amplification will result. In this condition, the power from each device will be fully absorbed into the RF load, and the overall efficiency will be equal to the efficiency of either PA. If the phase shift is adjusted to zero, then *in principle* the outputs from each device will phase cancel, and there will be no power dissipated in the RF load. In practice this cancellation may not be

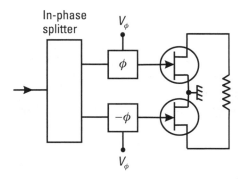

In-phase splitter

V_ϕ

ϕ

$-\phi$

V_ϕ

Figure 2.19 Basic Chireix outphasing PA configuration (biasing components not shown).

perfect, and some quantitative assessment is required to determine the precision of phase control needed to achieve a given dynamic range. There is also a more pragmatic concern about the likelihood of oscillation, and the ability of the devices to withstand the near open-circuit loads into which they must surely be working in this condition. On the basis of such end-point reasoning, it is clear that as the phase angle is adjusted from the anti-phase to the in-phase condition, the RF loading at each device goes through a dramatic change. It would seem appropriate to examine these functional dependencies on the angle ϕ.

The first of these dependencies is the relationship of the phase offset to the output amplitude, the basic process by which AM is "constructed." It is an important assumption in the Chireix system that each device is a saturated amplifier which generates a sinusoidal RF voltage having an amplitude V, where V is closely related, or approximates to, the dc supply rail voltage. So the voltage appearing across the RF load resistor is

$$\begin{aligned} v_o &= V\cos(\theta - \phi) - V\cos(\theta + \phi) \\ &= 2V\sin(\theta)\sin(\phi) \end{aligned} \tag{2.12}$$

Clearly, if it were possible to construct the phaseshifters such that the phase ϕ had an inverse sine drive characteristic,

$$\phi = k\sin^{-1}\left(V_\phi\right)$$

where V_ϕ is the drive signal applied to the phaseshifter, then linear changes in the drive signal V_ϕ would produce linear changes in the PA output amplitude. Chireix, and some others since, seem to make rather a lot out of this symbolic trick; in times of old, where such functions had to be created using analog circuitry, this may have been a helpful step in the process of defining suitable drive circuitry. A modern implementation would, of course, bypass such reasoning and apply the necessary phase control using a DSP look-up table. The issue then becomes one of precision. Figure 2.20 shows a logarithmic amplitude plot, in decibels versus outphasing angle, for the more critical low end of the dynamic range. The plot essentially speaks for itself; clearly the impact of requiring a 1° outphasing angle to achieve a 40-dB dynamic range of amplitude modulation depends on the system requirements. For an AM short wave transmitter, a legal and good fidelity signal could be constructed using maybe only 20 dB of dynamic range; AM transmitters do not have a requirement for maximum modulation "depth." This was probably the application that Chireix initially had in mind, and is in stark contrast to the dynamic range requirements of a modern communications system. It is clear that outphasing angle precision is a major problem.

The second dependency is the load-pulling effect; the impedance viewed by each device as the outphasing angle varies. Figure 2.21 shows the circuit typically assumed for this analysis, both by Chireix and others [4, 6]. Once again, the critical and idealizing assumption is made that each device is in a state of heavy rail clipping and as such can be considered to be a RF voltage source. The very limited validity of this assumption turns out to be a

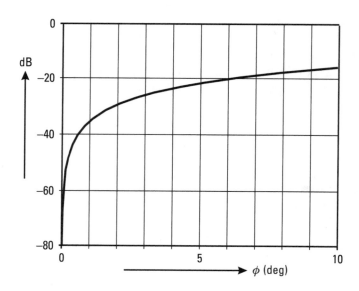

Figure 2.20 Outphasing cancellation.

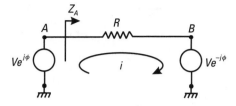

Figure 2.21 Schematic for Chireix analysis.

significant flaw in the realization of the attractive theoretical possibilities it creates. The circulating current in Figure 2.21 is given by

$$i = \frac{Ve^{j\phi} - Ve^{-j\phi}}{R}$$

so that the impedance presented to device "A" is given by

$$Z_A = \frac{Ve^{j\phi}}{i}$$

$$= \left(\frac{R}{2}\right)(1 - j\cot\phi)$$

and performing a series to parallel transformation, the impedance Z_A can be represented as a shunt resistance R_p, where

$$R_p = \left(\frac{R}{2}\right)\left(\frac{1}{\sin^2\phi}\right) \tag{2.13a}$$

and a shunt reactance X_p, where

$$X_p = \left(\frac{R}{2}\right)\left(\frac{-2}{\sin 2\phi}\right) \tag{2.13b}$$

These impedances are plotted, with values normalized to $R/2$, in Figure 2.22. Also included in Figure 2.22 is the power backoff, in decibels, as a function of outphasing angle.

These plots are worthy of close scrutiny, since they represent a major practical drawback of the Chireix method. They go some way towards

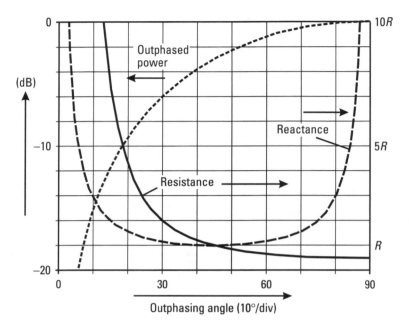

Figure 2.22 Outphasing impedance shift.

explaining why attempts to realize this kind of PA configuration can run into problems. Essentially, almost as soon as any outphasing action is attempted, the load impedance becomes highly reactive. A 45° value for ϕ gives only 3-dB power reduction, yet the reactive component already equals the real component. Beyond this point, even modest power-down values result in a widely varying range of almost entirely reactive impedances.

2.3.2 Discussion, Analysis, and Simulation

It is a matter of practical experience that RF power transistors do not "like" reactive loads. The plots in Figure 2.22 show that each device in a Chireix configuration will experience mainly reactive loads, beyond about the 3-dB backoff point. This problem is not a just simple issue of device stability and ruggedness. The reactive loads pose a much bigger question as to whether the saturating device still behaves like a voltage source. But if it were somehow possible to resonate out the reactive component of the outphasing imped-ance, the system would have some attractive possibilities. This is illustrated, through an actual simulation, in the waveforms shown in Figure 2.23. This shows a device operating in "Class FD" mode, as described in *RFPA*

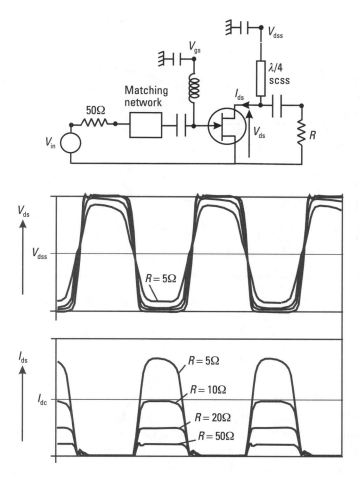

Figure 2.23 Simulation of "Class FD" PA using low parasitic device; once hard clipping starts, the device current becomes mainly a function of the RF load value (same input drive level in each case).

(Chapter 4). As the load resistor is increased in value, the voltage waveform remains essentially a rail-to-rail square wave, and the peak current drops. So increasing the load resistor reduces the RF power, but simultaneously the mean current will drop as well, thus maintaining good efficiency, as shown in Figure 2.24. This shows the intended action of the Chireix combiner, as the configuration is sometimes called. Unfortunately, the simultaneous presence of a dominating reactive component significantly changes this rather clean performance.

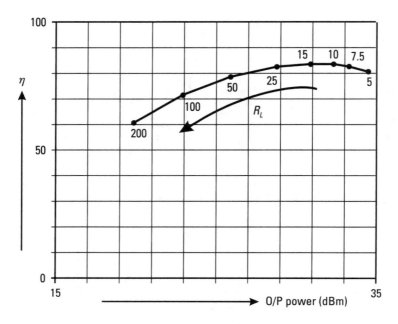

Figure 2.24 Simulated "Class FD" PA, showing effect of varying load resistance.

Chireix proposed the use of an additional shunt reactance across each device in order to resonate the out-phasing reactance at a single chosen value of ϕ. Figure 2.25 shows an impedance plot with two values of ϕ chosen for the resonance point. It can be seen from these plots that the resonance can only be made effective over a fairly narrow range of ϕ; the efficiency would be expected to show a sharp peak coinciding with the resonance. We will, however, refrain from reproducing the corresponding PBO efficiency analysis given by Chireix, and others [4, 6]. In a practical circuit, the efficiency is degraded by other effects, and the utility of the compensating reactance is questionable. But before talking realistically about efficiency, the voltage source assumption must be cast aside in favor of a full simulation using real device models. It will be seen that the reactive load and the finite "on" resistance of each device conspire further to degrade the performance of this configuration.

Figure 2.26 shows the schematic used for a simulation. A "real" model for the transistor is used, in fact the same GaAs MESFET device which was used extensively in *RFPA* for simulation of Class AB circuits.[1] It should be

---

1. See Appendix for model details.

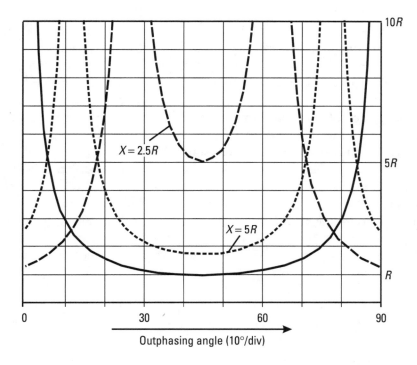

Figure 2.25 Effect of compensating shunt reactance on outphasing impedance (uncompensated reactive component shown, solid line).

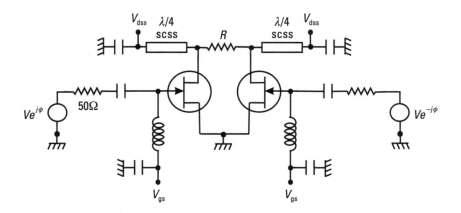

Figure 2.26 Schematic for simulation of Chireix outphasing circuit.

fairly noted that this particular device has very low parasitic reactances for the simulation frequency used (850 MHz). It also has plenty of gain, so no actual input matching is shown. This is a realistic situation, and an advantageous one, when using technologies such as GaAs MESFET, HBT, or Silicon Carbide MESFET at frequencies below 2 GHz. Unfortunately, the comparatively large parasitics in a Laterally Diffused Metal Oxide Semiconductor (LDMOS) device reduce the scope for success in these kinds of application. For simulation purposes, and clarity, the circuit elements have been kept to a minimum. In particular, the RF load is shown as a physical resistor of suitable value, rather than a 50-Ω termination and a matching network. The key point to note is that the load resistor will not, in general, have a virtual ground at its midpoint; the action required in this circuit is to force the circulating current to flow through both devices. This has some important implications for the realization of the necessary balun to convert the output into an unbalanced (coaxial) signal; this issue will be discussed in due course. The intent is to maintain Class FD operation, with each device hard-clipping on the supply rails (see *RFPA*, Chapter 5), so that the current at each device is mainly a function of the driving point impedance. The critical element to achieve this is the even harmonic short circuit, realized here as a thoroughly practical quarter-wave short circuit stub (SCSS). The Chireix compensating reactances will also be in shunt with each device.

Figure 2.27 shows a set of simulation results, initially with no compensation reactance. The immediate conclusion is that it basically seems to work. The action predicted by the idealized analysis can, in a quantitative sense, be clearly seen. As the outphasing angle ϕ is reduced from the anti-phase value of 90°, the RF output power goes down, but so does the dc supply, giving a major improvement in PBO efficiency. In fact, a more detailed examination of the waveforms in these simulations would seem to indicate that the physical action is not exactly as explained by the simple model using voltage generators. This does not affect the validity or usefulness of the simulation, but seems worthy of further study. Figure 2.28 shows some power-efficiency plots, including the effect of adding compensation reactance. Here we see a substantial problem; higher efficiency can be obtained, but this trades severely with dynamic range. Clearly, the asymmetry of the circuit interferes with the cancellation process. It is possible that this problem can be reduced by suitable offset of the differential phase settings, but the simulation seems to indicate that there is a fundamental tradeoff issue here. Once again, it is important to recall that old analog AM transmitters did not necessarily require *deep* AM to be applied to the carrier; fidelity could be maintained over a very shallow modulation depth, provided that the signal would always be received at a strong level. This is an assumption that can usually be made in commercial broadcasting. Modern high-density communications systems

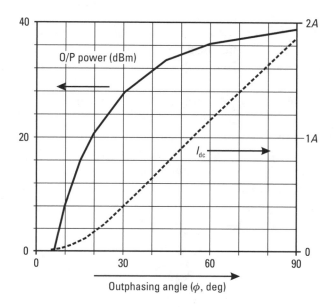

Figure 2.27 Outphasing PA simulation results; no reactive compensation.

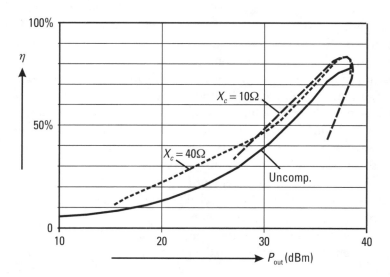

Figure 2.28 Power/efficiency plots for Chireix simulation; effect of compensating reactance included.

require much more dynamic range for the AM component, and this aspect of the Chireix system emerges as the most problematic. Nevertheless, the efficiency plots of Figure 2.28 compare very favorably with any other technique for envelope efficiency management, despite the fact that the system requires no additional current-hog such as the power converter in a Khan system.

The circuit used for this simulation is still somewhat idealized. One aspect which should be discussed is the need for a differential RF load. It is fair to speculate that the use of an inappropriate balun may have contributed to the lack of success in attempts to realize a circuit of this kind. The key issue is that the balun must float both input connections from the ground connection, and not place a ground at the midpoint of the resistor. This can be seen by reconsidering the schematic, Figure 2.18, and the analysis leading to (2.13). The voltage measured from the ground connection to the center point of the load resistor is given by

$$
\begin{aligned}
V_c &= \frac{V}{2}\left\{e^{j\phi} + e^{-j\phi}\right\} \\
&= V\cos\phi
\end{aligned}
$$

so that this point becomes a virtual ground only when $\phi = 90°$, the conventional push-pull condition. A push-pull amplifier can tolerate, and indeed may benefit, from a physical ground connection at this point. For example, at microwave frequencies it is common practice to use a power splitter and phaseshifter in order to make a quasi-balun, which is in effect just a power splitter having a 180° phase shift on one arm. Such a device would completely remove the load-pulling action by introducing high isolation between the two combining ports. It is an irony that all of the standard designs for microwave power combiners have been devised with a goal of maximizing such isolation between the combining devices.

Unfortunately, the requirement for a "true" balun whose inputs completely float above ground is quite challenging in a microwave context. The most promising approach is the balun structure shown in Figure 2.29. The outer of the coaxial section forms a quarter-wave SCSS and therefore presents an open circuit to ground at the input port. The inner section is completely shielded from ground and so the net effect, at the center frequency, approximates to the required balun action. The bandwidth of such an arrangement is strongly dependent on the characteristic impedance of the airline formed between the cable sheath and the ground plane. Increasing the spacing, or decreasing the cable diameter both lead to physical manufacturing and electromagnetic discontinuity problems, but a compromise can usually be found for bandwidths up to about 10%. It should be noted that since

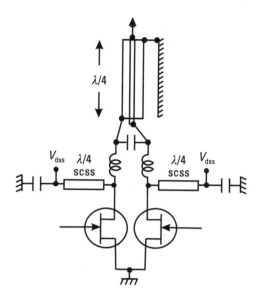

Figure 2.29 Possible realization of output matching and balun for a Chireix combiner.

the balun will probably be realized at the 50-Ω level, a balanced matching network will be required, as also shown in Figure 2.29. This is actually a good deal more convenient in its practical realization than the ubiquitous unbalanced networks, the ground points of which can introduce unwanted parasitic inductance.

This simulation indicates that this technique is viable for applications which require less than about 30-dB dynamic range in the AM component. It seems, however, that no such device has ever been built in the microwave frequency range.

2.3.3 Variations

The Chireix technique has one principal variation, which seems to have been the focus for most of the published literature on the subject. Essentially, this variation is to use the basic outphasing action to construct an AM signal, but to use a simple power combiner rather than a Chireix common load connection. Thus the key load-pulling action of the Chireix configuration is discarded. Such a configuration, shown in Figure 2.30, will offer no efficiency management; at low points of the envelope amplitude the power from the two PAs will be largely wasted, and dissipated in the "dump" port of the combiner. Thus the PBO efficiency curve shows a simple linear inverse

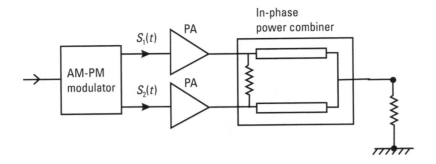

Figure 2.30 Outphasing system using conventional power combiner.

relationship with the power, shown in Figure 2.31. For example, even for a PA giving 80% efficiency at maximum output, the efficiency at the 10-dB turndown point will be 8%. This is considerably worse PBO efficiency performance than that which can be obtained from a Class AB amplifier. On the other hand, the configuration offers a clean approach for accurate and highly linear control of the AM using a suitable DSP driver. In this respect, direct comparison with a Class AB PA is inappropriate, due to the much higher potential linearity of the DSP controlled outphasing system. Such an AM control method would not require the added complexity and power wastage

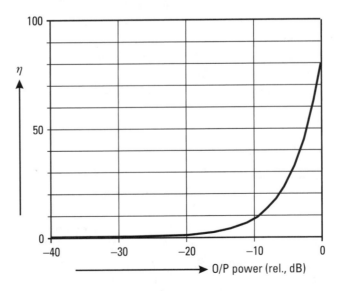

Figure 2.31 Efficiency/PBO characteristic for outphasing amplifier using a conventional power combiner (maximum efficiency of 80% assumed).

of the power converter required in a Khan system. It has even been proposed that the power dissipated in the combiner could be retrieved using an RF detection scheme [7].

Another variation is a hybrid arrangement using both Khan and outphasing methods together. This would be a potential way of solving the outstanding weakness of a simple Khan transmitter, which is the limited dynamic range. One possibility would be to have a number of fixed-supply voltages available. Outphasing could be used at each supply voltage to give, say, a 6-dB AM control range, and then the supply would be switched to the next voltage level, where the same range of outphasing would be repeated to obtain another 6-dB range with essentially the same efficiency. Figure 2.32 illustrates, in principle, the general form of PBO performance that should be possible. The key point here is that a limited number of fixed-voltage supplies could be generated at very high efficiency using standard switching power supply design methods. Figure 2.32 shows that the maximum efficiency will probably drop at each downward supply voltage step.

2.3.4 Chireix: Conclusions

Under the scrutiny of a modern CAD simulator, the Chireix outphasing technique appears to be capable of useful performance. Questions remain,

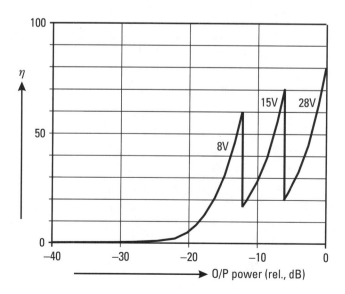

Figure 2.32 Conceptual outphasing system using three switched voltage supplies; outphasing is used to control power between supply voltage steps.

however, about the validity of the simple model which is frequently used to describe its action. The key element, which may have been missed by some workers in attempting experimental evaluation, is a floating common RF load between the devices. This entails unorthodox balun and matching network design in order to force the fundamental current component to flow through both devices. The kind of amplifier mode is also important; in order to approximate the necessary RF voltage source, an amplifier design is required in which the device is hard-clipping under all conditions. The downside of the technique is that limited dynamic range is available, especially if reactive compensation is used. Reactive compensation, a key element proposed originally by Chireix, does appear to have a potential role in improving the PBO efficiency characteristic. Even without reactive compensation, the PBO efficiency characteristic is attractive and very competitive in comparison to other techniques under this general heading.

Simple outphasing, using a power combiner and DSP phase control, may still offer some potential for using nonlinear PAs in linear applications (LINC). In all cases, however, these methods are transmitter techniques which require prior knowledge of the signal. In this sense they should not be compared directly with amplifier techniques.

References

[1]   Khan, L. R., "Single Sideband Transmission by Envelope Elimination and Restoration," *Proc. IRE*, Vol. 40, July 1952, pp. 803–806.

[2]   Heimbach, M., "Digital Multimode Technology Redefines the Nature of RF Transmission," *Appl. Microw. & Wireless*, August 2001.

[3]   Doherty, W. H., "A New High Efficiency Power Amplifier for Modulated Waves," *Proc. IRE*, Vol. 24, No. 9, September 1936, pp. 1163–1182.

[4]   Chireix, H., "High Power Outphasing Modulation," *Proc. IRE*, Vol. 23, No. 11, November 1935, pp. 1370–1392.

[5]   Raab, F. H., "Efficiency of Doherty RF Power Amplifier Systems," *IEEE Trans. on Broadcasting*, Vol. BC-33, No. 3, September 1987, pp. 77–83.

[6]   Raab, F. H., "Efficiency of Outphasing RF Power Amplifier Systems," *IEEE Trans. on Communications*, Vol. 33, No. 10, October 1985, pp. 1094–1099.

[7]   Langridge, R., et al., "A Power Reuse Technique for Improved Efficiency of Outphasing Microwave Power Amplifiers," *IEEE Trans. on Microwave Theory and Technology*, Vol. MTT-47, No. 8, August 1999, pp. 1467–1471.

3

Some Topics in PA Nonlinearity

3.1 Introduction

The modeling and simulation of PA nonlinearities, and in particular the impact these nonlinearities have in a modern digital communications system, are a current topic of intensive research worldwide. It is the subject of entire books and symposia, which on closer study seem frequently to conclude that more research is needed before even quite simple questions can be answered. It should therefore be stated clearly at the outset that this chapter does not set out to provide all the answers, and in particular does not attempt to cover the subject of modern digital modulation formats in an analytical or tutorial manner. What is presented here is a short list of relevant topics, some of a tutorial nature and others which may provide some new, or different, insight into certain aspects of a very challenging and difficult subject.

The main focus is on the so-called envelope simulation approach to nonlinear system simulation, and the associated PA models. These models are typically very different from those used to simulate the PA itself at component level, representing a higher level of abstraction. There is also a more general issue of whether polynomial functions are even appropriate for modeling the nonlinear characteristics of RFPAs. Effects such as asymmetrical IM and ACP spectral response, for example, cannot be predicted using the Volterra formulation in its standard form. An experimental technique will be described for evaluating, and characterizing, such asymmetrical effects. Finally, the central issue of high peak-to-average power signals and the impact of clipping their "infrequent" peaks will be discussed.

3.2   A Problem, a Solution, and Problems with the Solution

Figure 3.1 shows a magnitude and phase plot of a typical digital communications signal. Depending on the modulation system, such diagrams change in their detailed form, but in general several obvious comments can be made:

- The peak power is substantially higher than the mean power.
- The peak power is reached infrequently.
- The amplitude may, in some cases, be constrained never to reach zero (not actually in the example shown).
- The time trajectory is clearly driven by a distinct "beat," or symbol clock.
- The symbol clock rate is at least two orders of magnitude slower than the underlying RF carrier.
- There are two "time domains": the time domain in which the individual RF carrier variations can be resolved, and the "envelope" domain, where the modulation of the carrier can be resolved.

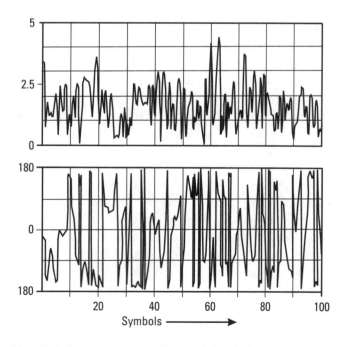

Figure 3.1 Magnitude (upper, arbitrary voltage units) and phase (lower, degrees) trajectories for a WCDMA signal.

- Distortion of any kind, either in the amplitude or in the phase, will result in spectral spreading of the signal.

There are also some less obvious comments:

- Although the transmitted signal follows a continuous trajectory, the demodulation process occurs at specific "constellation points."
- Distortion of the signal may not necessarily degrade the demodulation process.
- There is a *third* time domain of importance, the "measurement" domain, which represents the timescale on which the various measurement instruments accumulate, or average their readings.

It is the difference, of maybe six orders of magnitude, between the RF time domain and the measurement time domain, which poses a formidable problem in attempting to simulate a measured system response to signals of this kind. In order literally to simulate the task performed by a spectrum analyzer sweeping at a 1-kHz bandwidth, a 2-GHz RF system simulator has to run for at least 10^6 RF cycles, each of which has to be further subdivided into suitably small time samples. Papers in the literature which claim to simulate the ACP response of an RFPA admit, in some cases, to require hours of the fastest computation time. Yet an actual measurement takes only a matter of milliseconds. One could speculate that here lies an old divide between digital and analog computation, but this is not a favored viewpoint in the digital era.

In attempting to reduce the computation time to more manageable levels, a number of shortcuts have been devised. Each of these involves either simplifying physical assumptions, or the rejection of "redundant" information. Either way, caution must be exercised in placing too much reliance on the results. The most popular simplification, which has been the main simulation tool used over the last decade or so, is the so-called envelope domain simulator. This approach typically makes at least three major assumptions:

- The system is *quasi-static* in its response; the amplitude and phase of the output signal are the same for a given input RF carrier amplitude, regardless of how quickly they change, and are independent of previous "history."
- The timescale on which perceptible changes in the RF envelope occur is very slow in comparison to the RF time domain; this

implies that signals have "narrow" bandwidths when viewed in the RF spectral domain.

- The simulation bandwidth is restricted, by suitable filtering, to the immediate vicinity of the signal itself.

With these assumptions, the system can in principle be simulated following the process illustrated in Figure 3.2, and summarized as follows:

1. The AM-AM and AM-PM response of the PA is measured, on an *a priori* basis, using a cw signal source having a swept power range equivalent to the dynamic range of the desired signal.

2. The desired input signal is "created" in the envelope time domain.

3. At suitably closely sampled times in the envelope domain, the instantaneous amplitude and phase of the signal are determined from (2).

4. The AM-AM and AM-PM response is used, either in the look-up table (LUT) or in algorithmic form, to supply the gain compression and AM-PM at the current signal amplitude.

5. The output signal response at this instant of envelope domain time is constructed, using the gain and phase shift obtained in (4).

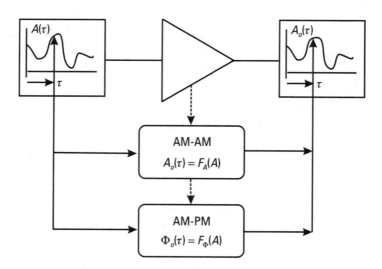

Figure 3.2 Envelope simulation concept.

6. The resulting output signal can be processed to obtain whatever information is required; this will usually include error vector magnitude (EVM), and fast Fourier transform (FFT), adjacent channel power (ACP), and mean power.

Note that the process generates both the amplitude and phase of the output signal as a function of envelope domain time. Clearly, the time samples have to be sufficiently close in order to obtain the desired spectral integrity.

The above process is appealing, inasmuch as it can be implemented by an RF engineer using measured PA data and a PC running a commercial math solver [1]. It also forms the basis for several commercial system simulation packages. There was a time, perhaps somewhere in the mid-1990s, when there was an uncomfortably large consensus that this was a true simulation process, having the same unimpeachable "correctness" as, say, the frequency response of a filter computed on a linear RF simulator. Unfortunately, the underlying assumptions relegate this technique to the ranks of "approximate" methodology: simple, understandable, intuitive, and most certainly useful—but not the final answer.

The biggest problem concerns the quasi-static assumption in the response of the RFPA. There are also some more subtle problems associated with the band-limited assumption that is implied in (2) above. The most tangible, and very troublesome, practical foundering of both assumptions is the inability of the envelope simulator to predict asymmetry in the IM or ACP response of an amplifier. This will be discussed in Section 3.5, along with some experimental data. Section 3.3 covers some important background material which has a wider relevance to nonlinear modeling methods.

3.3 Power Series, Volterra Series, and Polynomials

In times of old, the "power series" ruled the nonlinear RF world. Two-tone third-order intermodulation products (IM3s) had 3:1 slopes and intercepted the single tone linear response at a power level 9.6 dB higher than the 1-dB compression point. The reason, we were told in our youth by a pious senior member of technical staff, is that these effects are caused by third-degree nonlinearities. A surreptitious reference to the nearest available source containing the appropriate trigonometric formulae confirmed that the expansion of $(\cos\omega_1 t + \cos\omega_2 t)^3$ does indeed include terms in $\cos(2\omega_2 - \omega_1)t$ and $\cos(2\omega_1 - \omega_2)t$. Unfortunately, such references will usually not stretch to a fifth-degree expansion, and the vital recognition that IM3s can be generated also by higher degrees of nonlinearity than the third can be missed.

In fact, in the world of receivers, the simple third-degree results are accurate enough and are still widely used. A receiver in normal operation has signal excitations which are so small in comparison to the standing current and voltage bias points of the various active devices that third-degree effects dominate all of the nonlinear behavior. The intercept point and even the 1-dB compression point are just extrapolations, it being inconceivable that such signal levels would ever be reached in practice in a receiver application.

Clearly, this picture changes with an RF power amplifier. The 1-dB compression point may well lie inside the operating power range, and higher-degree nonlinear effects not only become significant, but can even dominate the nonlinear behavior. AM-PM effects start to become another important source of spectral distortion and demodulation errors. The IM power backoff (PBO) curves show more complex behavior, which may include null points and "plateaux." The issue of whether the traditional approach can be extended to handle this more complex situation has developed, in recent years, into something of a factional debate. The factions can be roughly categorized as follows:

(a) The traditionalists, who look no further than two-carrier testing, and still believe in constant IM slopes and intercept points;

(b) The neo-traditionalists, who believe that PAs can still be usefully, if not exactly, modeled using a Volterra, rather than a power, series which includes some higher-degree polynomial terms;

(c) The radicals, who assert that polynomial formulation breaks down when attempting to model something as nonlinear as an RFPA, especially Class AB PAs, and that either alternative methods need to be developed, or analytical treatment must retire and let computational number-crunching take over.

In this book, we take the neo-traditionalist position, (b). In radio frequency applications, there is fundamental justification for staying with a polynomial approach. The frequency domain, with its sinewave generators, bandpass filters, and spectrum analyzers, gives integral polynomial powers and coefficients tangible and measurable reality. The fact that some kinds of device may have characteristics that are more readily modeled by some other mathematical function is only of intermediate use if the final characteristic is to be transformed into the frequency domain; the FFT process itself infers a set of polynomial coefficients in the determination of harmonic frequency components. There are, nevertheless, some limitations to this approach. In particular, amplifiers which have substantial nonlinearity at the low end of their power range as well as the high end do pose serious problems. Class B, or deep Class AB PAs, clearly can exhibit this behavior to some extent.

While recognizing that this intermediate position will run into problems with some amplifier types, it should also be acknowledged that the useful range of the technique is highly dependent on the manner in which the polynomial coefficients, and the truncation of the power series, are derived. One of the proposals in this chapter is that more focused effort should be directed at using two-carrier tests to derive polynomial models rather than single-carrier gain and phase sweeps. A simple two-carrier test at the peak power level of a PA can give a firm indication as to how many degrees of distortion need to be included in the model. PBO sweeps of the various orders of IM can then be used to deduce a power series, and if phase information is available, the full Volterra series can, in principle, also be determined. In order to address this more challenging situation, an important first step is to recognize that the power series coefficients need to be vectors. This is the Volterra series formulation [2–5], which is now briefly reviewed.

Figure 3.3(a) shows a typical output spectrum for a PA, with a two-carrier input signal. The spectrum represents, approximately, a well-designed amplifier running around the 1-dB compression point. Clearly, there are third- and fifth-order IM products. Figure 3.3(b) shows a more conceptual plot, in which the individual spectral outputs are broken into separate components, each coming from a different degree of nonlinearity. For example, the third-order IM product has two components, one from the third-degree nonlinearity and another from the fifth degree. Each fundamental output has three components: the linear term, and a third- and a fifth-degree nonlinear contribution. All of these spectral components can be characterized by a simple power series for the amplifier,

$$v_o(t) = a_1 v_i(t) + a_3 v_i^3(t) + a_5 v_i^5(t) \tag{3.1}$$

Note that because it is assumed that the amplifier has a narrow fractional bandwidth, greater but comparable to the carrier spacing, only odd-order nonlinearities will generate in-band distortion products. This particular simplification can cause justifiable consternation, for example, in a Class AB amplifier which relies heavily on second-degree effects for its operation. This is where it is important to recognize that in this model we are working at the

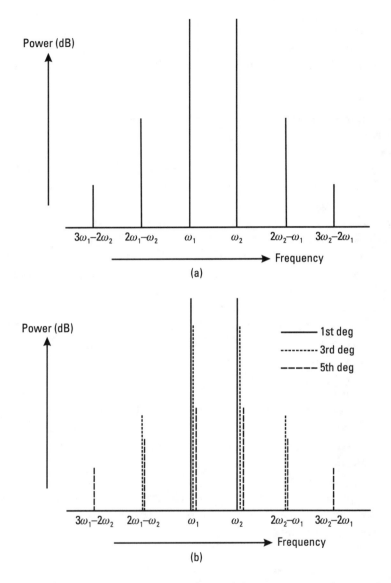

Figure 3.3 (a) Typical two-carrier IM response and (b) response showing separate components of IMs from different degrees of nonlinearity.

highest level of abstraction, at least one and maybe two levels higher than that which is typically used when designing the amplifier circuits. Equation (3.1) is entirely "behavioral" and is simply offered as a convenient formulation for fitting an *a priori* set of characterization measurements. It may even happen that within the amplifier, second-degree effects may themselves affect the level of third-degree distortion products. This was discussed in *RFPA* (Chapter 4), and is a direct consequence of inadequate harmonic termination. This, however, in no way violates the ability of (3.1) to model the final terminal characteristics of the amplifier.

In fact, there is no fundamental objection in including even power terms in (3.1), if this appears to improve the fit between the model and the measured data. But such terms will be "inert" as far as computations for close-to-carrier IM and ACP distortion are concerned. There are, however, some flaws in the simple formulation of (3.1). If even-power nonlinear processes are present within the device, there will be variations in the "dc" conditions.[1] In particular, a device which has significant even-degree distortion will show a low-frequency ac component on its "dc" supply. If this low-frequency component is allowed to interact in any way with the rest of the circuit, there may be some effects on the close-to-carrier odd-degree distortion which the simple formulation of (3.1) is not able to model. Such effects include asymmetry in the IM upper and lower sidebands, and will form the subject of Section 3.5. A more immediate and serious flaw, however, is the inability of (3.1) to model AM-PM effects. This can be resolved using a more general form of (3.1) proposed by Volterra [2].

For an input signal having the form

$$v_i(t) = v \cos \omega_1 t + v \cos \omega_2 t$$

it is clear that the third- and fifth-degree expansions of (3.1) will contain fundamental terms which are proportional to v^3 and v^5, respectively. There will also be terms at the third-order IM frequencies, $(2\omega_1 - \omega_2)$ and $(2\omega_2 - \omega_1)$, proportional to v^3 and v^5. Specifically, at each fundamental, the output is

$$v_{ofund} = \left\{ a_1 v + \frac{9}{4} a_3 v^3 + \frac{25}{4} a_5 v^5 \right\} \cos\left(\omega_{1,2} t\right)$$

and at a third-order intermodulation (IM3) frequency, the output is

$$v_{oIM3} = \left\{ \frac{3}{4} a_3 v^3 + \frac{25}{8} a_5 v^5 \right\} \cos\left(2\omega_{1,2} - \omega_{2,1}\right)$$

1. Obviously, if the supply has an ac component, it isn't dc any more. For clarity, we compromise and put "dc" in quotations.

It is in the summation of the individual components at each frequency that the Volterra series takes a crucial step towards higher generality; each degree of nonlinearity is defined to have a characteristic phase angle, φ_1, φ_3, φ_5, respectively. These angles are additional parameters which characterize the system, and may have any value between $-\pi$ and π. The summation of the components now has to be performed vectorially, as illustrated for the fundamental case in Figure 3.4. Volterra's postulation was that weakly nonlinear systems could be thus represented, the key point being that the phase angles are constant parameters, whatever the excitation.[2] Figure 3.4 shows the simplest possible case, where only a third-degree nonlinearity is added, at its characteristic phase angle φ_3, to the fundamental. The resultant, through its phase angle φ_3 shows an increasing departure from the original linear phase (φ_1 is taken as the horizontal reference phase direction) as the drive level is increased; this is of course the measurable process of AM-PM, giving a phase shift angle ϕ at the specified drive level.

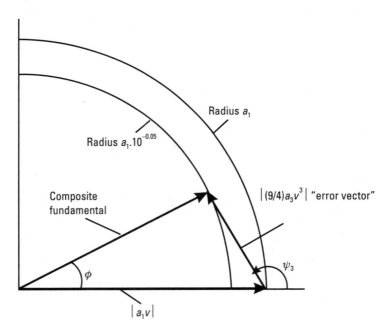

Figure 3.4 Vector addition of fundamental spectral components in two-carrier spectrum shown in Figure 3.3 (third-degree linearity only).

---

2. Volterra used a slightly more general formulation with a characteristic time delay for each degree of nonlinearity.

At a utilitarian level, the Volterra series is able to model the AM-AM and AM-PM distortion of a typical power amplifier, well into its compression region. An important issue is how many polynomial terms are required in a typical case. The widespread availability of polynomial curve-fitting routines in commercial mathematical software packages seems to have spawned something of a cavalier attitude in this area; "...might as well use fifty terms rather than five, the computer doesn't care..." would be a characterization of this phenomenon. There are however some severe hazards in creating curves using excessively high-degree polynomial functions. In particular, applications which will be using many derivative orders of the modeled curve require special care which is not always provided by the software package. We will return to this issue shortly; however, for the present time it is instructive to observe the kinds of PA characteristics which simple third-, or third- and fifth-, degree Volterra series can generate.

Figure 3.5 shows the simplest case of an amplifier with only third-degree distortion, driven up to its 2-dB compression point. The AM-AM response has a compression characteristic which is probably "softer" than a typical PA design; this can be resolved by using some higher-degree terms. The AM-PM response also shows a characteristic shape, with a scaling factor which relates to the value of φ_3. Imagining that three cases of φ_3 (180°, 160°,

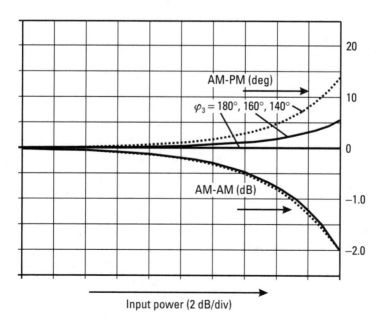

Figure 3.5 Third-degree PA distortion characteristics.

and 140°) correspond to separate physical amplifiers, it is important to recognize that although each amplifier displays the same AM-AM characteristic, *the a_3 coefficients will be different in each case.* This effect can be seen clearly by looking at the vector plot shown in Figure 3.4. At any given level of gain compression, the length of the "error vector," which in this case is just a third-degree term, will increase as the angle φ_3 departs from 180°. This means that the value of a_3 will be higher for the amplifiers which display higher AM-PM. Thus these amplifiers will also display higher levels of IM3, at a given drive level, since the IM3 terms are proportional to the a_3 value. In this sense, the contribution of AM-PM to the IM level in an amplifier can be quantified. It is important to note that in this formulation, the AM-AM and AM-PM distortion are *each* functions of *both* a_3 and φ_3. If, by some means, it were possible to take the amplifier with the highest AM-PM distortion (the case $\varphi_3 = 140°$) and physically remove or neutralize the AM-PM, then both of the Volterra coefficients, a_3 and φ_3, would have to be changed.

It is worth noting here in passing that the modern trend towards specifying power amplifiers using the error vector, and in particular its magnitude [error vector magnitude (EVM)] is well supported by the above analysis. The vector approach recognizes that AM-AM and AM-PM are both manifestations of a fundamental process of distortion which is not adequately described by the ubiquitous gain compression specification method.

Figure 3.6 shows a more representative PA gain compression and characteristic. This was generated using a Volterra series which contains third- and fifth-degree terms.

The fifth-degree term takes off in the vicinity of the 1-dB compression point and gives the characteristic more realistic sharpness. This added dimension of nonlinearity enables a much wider range of observed PA characteristics to be modeled, at least in terms of capturing their essential features. Figure 3.6 represents a very "well-behaved" characteristic, perhaps what might be expected from an optimized Class A design. In Chapter 1 it was shown that Class AB amplifiers can display a gain compression characteristic which shows a substantial range of low (less than 0.5 dB) compression, extending perhaps 6–10 dB down from the onset of saturation. Figure 3.7 shows a range of AM-AM characteristics, which can be modeled using suitably chosen values for a_3 and a_5, which show this feature. Figure 3.7 also shows the possibility of some mild gain expansion prior to the final saturation. This is quite a common practical observation, although the possible thermal origin for this behavior raises questions about the quasi-static assumption which underlies the use of a model of this kind.

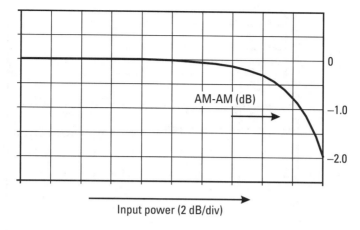

Figure 3.6 Fifth-degree AM-AM response.

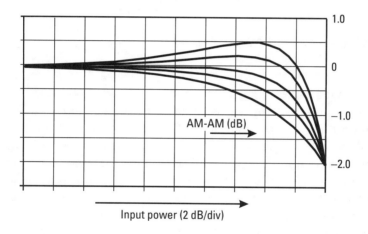

Figure 3.7 PA characteristics modeled using Volterra series having third- and fifth-degree terms.

Figure 3.8 shows a corresponding set of AM-PM curves; these can be obtained for essentially any of the AM-AM curves in Figure 3.6 by suitable choice of φ_3 and φ_5 coefficients. The three permutations of sign for φ_3 and φ_5 result in the curves shown in Figure 3.8; scaling of each parameter will produce greater or lesser overall magnitudes of AM-PM. The reversal of direction in AM-PM in the vicinity of the 1-dB compression point, caused by φ_3 and φ_5 having opposite signs, is also a common experimental observation.

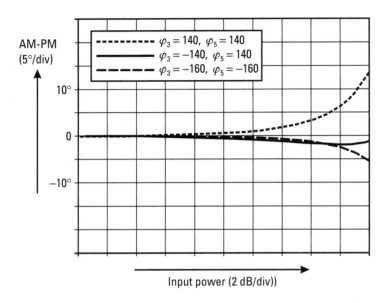

Figure 3.8 Fifth-degree AM-PM responses.

All of the curves in Figures 3.7 and 3.8 have been obtained by specifying a gain compression (or expansion) level at a chosen PBO level. Figure 3.7 was drawn using compression values of −0.5, −0.25, 0.05, 0.25, and 0.5 at the 6-dB backoff point, "backoff" being specified as measured from the 2-dB compression point in each case. So if these curves, and the underlying fifth-degree Volterra series equations, were being used to fit experimental measurements, only two power levels will be modeled precisely. Clearly, if higher degrees of nonlinearity are added, the number of precisely modeled points will increase in a corresponding manner.

This does not, however, guarantee that the model will do a better job. A simple, albeit possibly trivial, example of this is shown in Figure 3.9. This curve is generated using the same routine as in Figure 3.7, but an attempt was made to specify a higher compression level (1 dB) at the 10-dB backoff point in order to generate a softer characteristic. The two "exact" points are still modeled precisely, but the curve as a whole would obviously be useless as a model. This effect can actually be more dangerous when a higher-degree polynomial is used. This enables more "exact" points but opens up possibilities for some unrealistic gyrations between the fitted points. Some commercial curve-fitting routines recognize this problem and fit different polynomial functions to different parts of the data plot. The separate sections are blended together by forcing the derivatives to match up at the interface points. Such

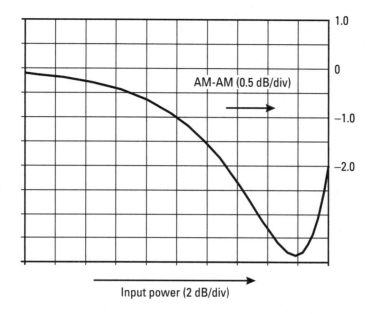

Figure 3.9 Example of potential hazards in using polynomial curve fitting to model PA characteristics (curve fitted using stipulated 1-dB and 2-dB compression points).

"spline curve" routines are useful for drawing a smooth line through a set of data points, but clearly do not give a closed-form analytical model. In general, a polynomial curve fitter for this application has to work on the data plot asymptotically, rather than on a basis of hitting individual points.

It is not clear whether all, or even any, of the polynomial curve-fitting routines which form part of commercial math software packages are sufficiently sympathetic to these special requirements. It is therefore necessary to re-examine the subject, using physical rather than numerical considerations. In particular, it would seem an intuitively reasonable assumption that if, say, the 51st-order IM products are of no physical or measurable significance for a given device under a given set of limiting conditions, a 51st-degree polynomial is an inappropriate model. So a good starting point in deciding the required degree of polynomial is a physical measurement, rather than an arbitrary numerical selection. One of the great advantages the PA modeler has is easy access to accurate experimental data on the derivatives of the device characteristic. This is, of course, the swept power IM data, which for a PA application is readily available at much higher levels of precision than the swept gain and phase characteristic.

The question of precision emerges as the core problem in using the AM-AM and AM-PM characteristics as the experimental data basis for determining the Volterra coefficients. This is illustrated in Figure 3.10. A PA gain compression characteristic is shown which has been generated using a power series with third-, fifth-, and seventh-degree components:

$$v_o = a_1 v + a_3 v^3 + a_5 v^5 + a_7 v^7$$

with $a_1 = 1$, $a_3 = -0.2$, $a_5 = 0.2$, $a_7 = -0.15$.

Figure 3.10 shows another gain curve, which at every point over the 20-dB PBO range lies within 0.05 dB of the actual curve. This second dotted curve has the parameters

$$a_1 = 1, \; a_3 = -.13, \; a_5 = -0.2, \; a_7 = 0$$

which will clearly yield a very different set of IM plots, as shown in Figure 3.11. Obviously, the "dotted" fifth-degree model will show no IM products of a higher order than 5. Although the IM3 plots are somewhat similar, there is clearly a large and easily measurable discrepancy in the IM5 plots. The gain plots, however, are clearly already drawn at the limit of resolution for a typical vector analyzer measurement. Although averaging techniques may be

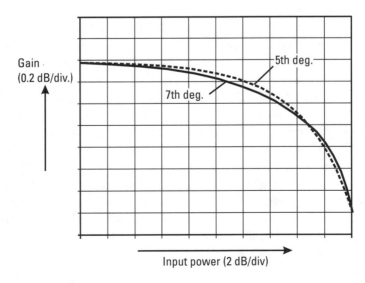

Figure 3.10 Fifth- and seventh-degree models, gain compression characteristic (note 0.2 dB/div gain scale).

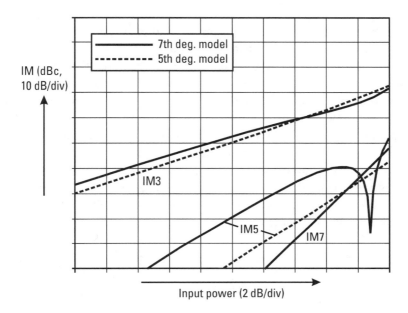

IM (dBc, 10 dB/div)

7th deg. model
5th deg. model

IM3

IM5

IM7

Input power (2 dB/div)

Figure 3.11 Fifth- and seventh-degree IM characteristics (gain responses shown in Figure 3.10).

used to improve the resolution, this would appear to be a step in the wrong direction for dynamic applications, which demand faster sweep speeds. It is apparent that two-carrier measurements offer a much sounder basis for determining Volterra coefficients than single-carrier gain and phase measurements. This forms the subject of Section 3.4.

3.4 Two-Carrier Characterization

The two-carrier, or "two-tone" test has a long history. Fundamentally, it is a convenient method of generating an amplitude-modulated carrier with essentially no distortion. Any attempt to create AM using some form of modulator runs into the problem that the signal will have distortion due to the nonlinearity of the modulator. In modern communications applications, the two-carrier test has been largely replaced by tests using the actual modulation system in use. Obviously, such testing is essential, both during development and in production, for determining specification compliance on a product. This section explores the concept that the two-carrier test still has much potential value as a *characterization* tool.

Returning to Figure 3.11, this simple example is a demonstration of the value of two-carrier IM plots in giving more direct, and more accurately measurable, information relating to the power series coefficients. Unfortunately, the IM amplitudes are only half of the story. Spectrum analyzers do not measure phase, so the hapless modeler is forced back to the single-carrier gain response, where phase information is available using a vector network analyzer. In fact, some modern network analyzers[3] can be configured to make IM phase measurements, and in any case it is not so difficult to conceive of a test setup which could do such a job. Figure 3.12 shows the result that such a measurement would produce, using the original seventh-degree device model, but with the phase angles changed to more realistic values so that AM-PM effects are included:

$$a_1 = 1, \ a_3 = 0.2, \ a_5 = 0.2, \ a_7 = 0.15,$$

$$\varphi_1 = 0 \text{ (relative)}, \ \varphi_3 = 150°, \ \varphi_5 = 10°, \ \varphi_7 = 170°$$

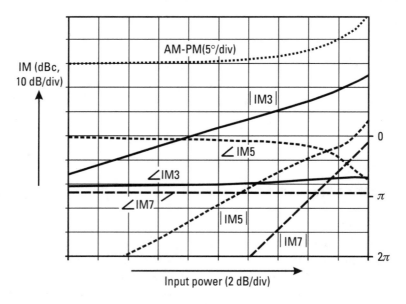

Figure 3.12 Seventh-degree IM characteristics, AM-PM effects included.

---

3. In particular, vector signal analyzers can usually be harnessed into making such measurements.

It seems clear that these plots contain the information required to derive a set of values for the Volterra coefficients, in a more directly accessible form than a simple gain and phase plot. First and foremost, the degree of polynomial truncation can be reasonably set to coincide with the highest order of IM product which has a measurable magnitude at the peak power level of the system. The Volterra angles, φ_n, can in principle be obtained directly from the asymptotic values they have at well backed-off levels. The magnitudes can then be modeled by scaling and combining the various individual IM slopes. It is important to note the part the Volterra angles play in shaping the IM characteristics. For example, comparing Figures 3.11 and 3.12, the original deep null in the IM5 plot is suppressed in Figure 3.11 due to the non-phase opposition of the φ_5 and φ_7 angles (150°, 10°). But the abrupt shift in the IM5 phase angle still clearly indicates where the IM5 becomes dominated by the seventh-degree nonlinearity.

Clearly, the derivation of the Volterra coefficients from the IM plots will involve a fitting algorithm that can be "seeded" with a rough guess based on simple graphical constructions. Fitting a polynomial characteristic to IM plots does require some additional mathematical homework, which is presented here since it may not have appeared anywhere else; the extension of the necessary trigonometric expansions up to powers higher than the third and fifth can be laborious and math solvers cannot always rise to the task.

Representing a two-carrier excitation in the form

$$v_{in} = v\cos(\theta_c - \theta_m) + v\cos(\theta_c + \theta_m)$$
$$= 2v\cos\theta_m\cos\theta_c$$

where $\theta_m = \omega_m t$ represents the amplitude modulation, and $\theta_c = \omega_c t$, the RF carrier.

The odd-degree power series for this input signal has the form

$$v_o = a_1 2v\left(\cos\theta_m\cos\theta_c\right) + a_3 2^3 v^3\left(\cos\theta_m\cos\theta_c\right)^3 + a_5 2^5\left(\cos\theta_m\cos\theta_c\right)^5$$
$$+ a_n 2^n v^n\left(\cos\theta_m\cos\theta_c\right)^n \tag{3.2}$$

so that the nth degree term can be written as

$$v_{on} = a_n 2^n v^n \cos^n\theta_m \cos^n\theta_c \tag{3.3}$$

The odd degree expansion of $\cos^n\theta$ can be obtained by application of De Moivre's theorem and results in a general expression

$$\cos^n \theta = \frac{1}{2^{n-1}} \left[\cos n\theta + n\cos(n-2)\theta + \frac{n!}{(n-k)!k!} \cos(n-2k)\theta \right],$$

$$1 < k < (n-1)/2 \tag{3.4}$$

In a narrow band-limited system, the nth degree output voltage contribution from (3.3) can be reduced to the modulation on the fundamental carrier,

$$v_{on} = a_n 2^n v^n \frac{n!}{2^{n-1} \left(\frac{n-1}{2} \right)! \left(\frac{n+1}{2} \right)!} \frac{1}{2^{n-1}}$$

$$\left\{ \cos n\theta_m + n\cos(n-2)\theta_m + \frac{n!}{(n-k)!k!} \cos(n-2k)\theta_m \right\} \cos\theta_c$$

$$\tag{3.5}$$

The key point about (3.5) is that each term in the bracket is the magnitude of the $IM_{(n-2k)}$ product component for the nth degree of nonlinearity. For example, for the third-degree case, $n = 3$, the term $\cos 3\theta_m \cos\theta_c$ represents the third-order intermodulation component IM_3, caused by the third-degree nonlinearity. So the magnitude of this component is

$$a_3 v^3 2^3 \frac{3!}{2^2} \frac{1!}{1!2!} \frac{1}{2^2} = \frac{3}{2} a_3 v3$$

and the magnitude IM3 components are equal to $\frac{3}{4} a_3 v^3$, from the expansion of $\frac{3}{2} a_3 v^3 \cos 3\theta_m \cos\theta_c$ into upper and lower sideband components.

Table 3.1 presents the results obtained from (3.5) for degrees of nonlinearity up to the ninth. (The fundamental gain compression components are, for convenience, labeled as "IM1.")

Figure 3.13 shows an example of a more complex characteristic, which uses a ninth-order power series [$a_1 = 1$ (0°), $a_3 = 0.1$ (0°), $a_5 = 0.35$ (180°), $a_7 = 0.5$ (0°), $a_9 = 0.35$ (180°)]. This yields a set of characteristics which are quite typical of higher power devices, such as LDMOS, which have been carefully biased and tuned in order to present favorable nulls in the IM characteristics. Such devices are notable for flat gain characteristics and a very abrupt compression. This requires higher order terms, up to the ninth in this case, but the four coefficients can be quite easily estimated from the IM curves. In practice the non-ideal values of the Volterra series phase angles will

Table 3.1
Power Series Coefficients

n	3	5	7	9
	a_3v^3	a_5v^5	a_7v^7	a_9v^9
IM1	$\dfrac{9}{4}$	$\dfrac{25}{4}$	$\dfrac{1225}{64}$	$\dfrac{3969}{64}$
IM3	$\dfrac{3}{4}$	$\dfrac{25}{8}$	$\dfrac{735}{64}$	$\dfrac{1323}{64}$
IM5	—	$\dfrac{5}{8}$	$\dfrac{245}{64}$	$\dfrac{567}{64}$
IM7	—	—	$\dfrac{35}{64}$	$\dfrac{567}{256}$
IM9	—	—	—	$\dfrac{3969}{256}$

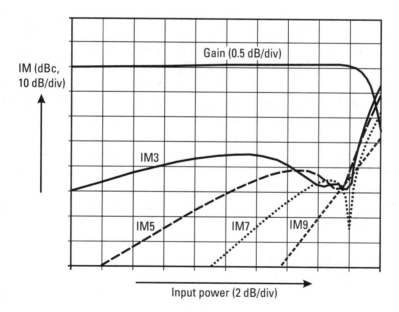

Figure 3.13 Ninth-degree gain and IM characteristic.

make the process less straightforward, but still quite feasible. The following procedure could be followed for modeling a device in this manner.

1. Start off with a power series and an amplitude IM plot.

2. Determine the highest order IM ("n") which is significant for characterization purposes, at the maximum peak power level (say, -50 dBc).

3. Assume the highest significant IM order has a simple n: 1-dB rolloff.

4. Determine the nth degree term from IMn level at maximum drive.

5. Fit successive lower IM plots using the (now determined) higher-degree terms and a chosen value of current degree IM; use just positive or negative values.

6. Using IM phase plot (if available), obtain Volterra phase angles [i.e., from $0°$ or $180°$ values used in (5)] and refine overall fit.

Two-carrier characterization has another vital advantage over continuous, or slow swept, measurements. The separation of the two RF carriers gives an AM "beat" cycle which can be set to a frequency appropriate to the final application. In a digital communications system, for example, the separation could be advantageously set to the symbol rate. In this manner, some of the "memory effects" which power amplifiers display can be partially absorbed in the characterization process. These effects form the subject of Section 3.5, but at this point we can note that, for example, when sweeping a power amplifier over a 20–30-dB range, it is inevitable that heating and cooling of the active devices will occur during the sweep. The impact of these effects can be very significant when measuring gain and phase variations at the precision level required in this form of characterization. In particular, thermal effects will be much more prominent in a slow sweep, or a stepped cw test. In a two-carrier test, the device can at least be measured in a comparable memory environment to that which will apply in the final application.

3.5 Memory and IM Asymmetry in RFPAs

It is a common practical observation that upper and lower IM sidebands are asymmetrical, typically in the order of a decibel or two on a typical spectrum analyzer display. The asymmetry is often dependent on the carrier spacing, but not in a monotonic fashion. This observation has frustrated at least one generation of microwave technicians and test engineers, not just because the asymmetry takes the product out of spec but also the unsatisfactory attempts to explain the cause of the effect. First and foremost, the effect can certainly be explained as an interaction between AM-AM and AM-PM distortion processes. On the other hand, the mere presence of both processes does not guarantee that asymmetry will occur.

Mathematical analysis of the problem is a little cumbersome, but in view of the major impact on practical PA applications it is a very worthwhile exercise. Figure 3.14 shows the critical difference which is required in the dynamic AM-AM and AM-PM characteristics in order for IM asymmetry to be observed. Essentially, if there is a time lag, or phase shift *as measured in the envelope time domain*, between the AM-AM and AM-PM responses, or their individual frequency components, IM asymmetry will occur. This is an entirely behavioral characterization of the problem; no statement is needed as to what physical mechanisms may cause such a response. For the purposes of simplification, without losing this critical generalization, the AM-PM phase response will be considered to be sinusoidal, but with an envelope domain phase shift of Δ with respect to the amplitude distortion. In practice, and as indicated in Figure 3.14, the measured dynamic AM-PM response will tend to show an asymmetrical characteristic which peaks at the PEP point. The fundamental component of such a waveform will display the phase shift Δ with respect to the AM-AM response.

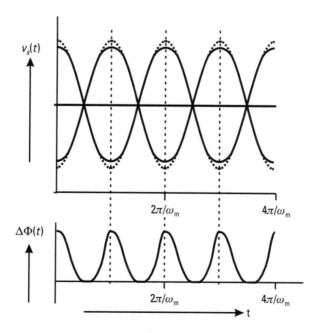

Figure 3.14 PA output envelope, AM-AM and AM-PM dynamic response; note the asymmetrical AM-PM characteristic.

With the above considerations and simplifications, the PA output signal in Figure 3.14 can be represented as

$$v_o(t) = \left[a_1 \cos(\Omega t + \Delta) + a_3 \cos 3(\Omega t + \Delta) \right]\left[\cos\{\omega t + \Phi\cos(2\Omega t)\}\right]$$

$$(3.6)$$

where Ω is the two-carrier beat frequency (or half of the carrier frequency separation), a_1 and a_3 are the power series amplitude distortion coefficients, Φ is the peak amplitude of the AM-PM distortion, and ω is the RF carrier frequency. Δ is the all-important envelope time domain phase angle between the occurrence of AM-AM and AM-PM effects; for analytical convenience this relative phase shift has been incorporated into the AM-AM distortion term in (3.6). The output is assumed, for convenience, to be taken for unity input amplitude so the first-order, third-degree voltage term coefficient $\left(\frac{3}{4}\right)a_3$ is absorbed into the chosen value of a_3. Note particularly that the AM-PM response has twice the frequency of the envelope modulation, since there are two envelope peaks for each cycle of Ω.

A final simplification can be made, which is to assume that the peak AM-PM phase angle Φ can be assumed small enough so that $\sin\Phi = \Phi$. This avoids the analysis escalating into Bessel functions, and does not impact the basic occurrence of IM asymmetry. The first two levels of expansion of (3.6) can be performed using well-known trigonometric identities:

$$
\begin{aligned}
v_o(t) &= \left[a_1 \cos(\Omega t + \Delta) + a_3 \cos(\Omega t + \Delta) \right]\left[\cos\{\omega t + \Phi\cos(2\Omega t)\}\right] \\
&= \left[a_1 \cos(\Omega t + \Delta) + a_3 \cos 3(\Omega t + \Delta) \right] \\
&\quad \left[\cos(\omega t)\cos(\Phi\cos(2\Omega t)) - \sin(\omega t)\sin(\Phi\cos(2\Omega t))\right] \\
&\approx \left[a_1 \cos(\Omega t + \Delta) + a_3 \cos 3(\Omega t + \Delta) \right]\left[\cos(\omega t) - \sin(\omega t)\Phi\cos(2\Omega t)\right]
\end{aligned}
$$

$$
\begin{aligned}
&= \left[a_1 \cos(\Omega t + \Delta) + a_3 \cos 3(\Omega t + \Delta) \right] \\
&\quad \left[\cos(\omega t) - \frac{\Phi}{2}\{\sin(\omega + 2\Omega)t + \sin(\omega - 2\Omega)t\}\right]
\end{aligned}
$$

$$(3.7)$$

Note that in (3.7), the third-order IMs, at frequencies $\omega \pm 3\Omega$, arise from two products, the AM-AM component coming from the product of the third-degree AM term in the first bracket and the RF carrier in the second bracket, and the AM-PM component which comes from the fundamental

AM term in the first bracket and the sinusoidal $\omega \pm 2\Omega$ terms in the second bracket. There is a key mathematical difference between these two products; in one case the RF carrier is sinusoidal, and in the other case it is cosinusoidal. This is very much the core of the analysis and requires much care in further expansion:

Taking first the AM component,

$$IM3_{AM} = \left[a_3 \cos 3(\Omega t + \Delta)\right] \cos \omega t$$

$$= \frac{a_3}{2} \left\{\cos(\omega t + 3\Omega t + 3\Delta) + \cos(\omega t - 3\Omega t - 3\Delta)\right\}$$

$$= \frac{a_3}{2} \left\{ \begin{array}{l} \cos 3\Delta \cos(\omega + 3\Omega)t - \sin(3\Delta)\sin(\omega + 3\Omega)t \\ + \cos(-3\Delta)\cos(\omega - 3\Omega)t - \sin(-3\Delta)\sin(\omega - 3\Omega)t \end{array} \right\}$$

$$= \frac{a_3}{2} \left\{ \begin{array}{l} \cos 3\Delta \cos(\omega + 3\Omega)t - \sin(3\Delta)\sin(\omega + 3\Omega)t \\ + \cos(3\Delta)\cos(\omega - 3\Omega)t + \sin(3\Delta)\sin(\omega - 3\Omega)t \end{array} \right\}$$

$$\text{(3.8)}$$

The AM-PM component at the IM3 frequencies comes from the second product,

$$\left[a_1 \cos(\Omega + \Delta)t\right]\left[\frac{\Phi}{2}\left\{\sin(\omega + 2\Omega)t + \sin(\omega + 2\Omega)t\right\}\right]$$

$$= \frac{a_1 \Phi}{4}\left\{\sin(\omega t + 3\Omega t + \Delta) + \sin(\omega t + \Omega t - \Delta)\right.$$

$$+ \sin(\omega t - \Omega t + \Delta) + \sin(\omega t - 3\Omega t - \Delta)\}$$

from which the IM3 components are

$$IM3_{PM} = \frac{a_1 \Phi}{4}\left\{\sin(\omega t + 3\Omega t + \Delta) + \sin(\omega t - 3\Omega t - \Delta)\right\}$$

$$= \frac{a_1 \Phi}{4}\left\{\cos \Delta \sin(\omega + 3\Omega)t + \sin(\Delta)\cos(\omega + 3\Omega)t \right. \quad \text{(3.9)}$$

$$+ \cos(\Delta)\sin(\omega - 3\Omega)t - \sin(\Delta)\cos(\omega - 3\Omega)t\}$$

So the combined upper IM3 sideband, at frequency $\omega_{3U} = \omega + 3\Omega$ is

$$IM3_{USB} = \left\{ \frac{a_3}{2} \cos 3\Delta + \frac{a_1 \Phi}{4} \sin \Delta \right\} \cos \omega_{3U}$$

$$+ \left\{ \frac{a_1 \Phi}{4} \cos \Delta - \frac{a_3}{2} \sin 3\Delta \right\} \sin \omega_{3U}$$

and the combined lower IM3 sideband at frequency $\omega_{3L} = \omega - 3\Omega$ is

$$IM3_{LSB} = \left\{ \frac{a_3}{2} \cos 3\Delta - \frac{a_1 \Phi}{4} \sin \Delta \right\} \cos \omega_{3L}$$

$$+ \left\{ \frac{a_1 \Phi}{4} \cos \Delta + \frac{a_3}{2} \sin 3\Delta \right\} \sin \omega_{3L}$$

$$(3.10)$$

Clearly, the upper and lower IM3 sidebands will display asymmetry for non-zero values of the envelope domain offset phase angle Δ.

Although simple enough in concept, formulation, and derivation, (3.10) would appear to qualify as a landmark result. Too frequently, the assertion that AM-PM in itself causes IM asymmetry has been the accepted explanation. This seems to arise from some misplaced comparisons between the AM-PM distortion of a two-carrier system and the standard textbook analysis for the intentional phase modulation of a single carrier. It is, however, still true to state that AM-PM effects cause the asymmetry, in the sense that if AM-PM were absent, there would be no asymmetry. So reduction of AM-PM in the PA design will, in the first place, alleviate the problem. The fact that a time delay between the physical processes of amplitude and phase distortion causes asymmetry in IM and ACP spectra sheds further light on this subject, however. Not only do physical explanations for the effect immediately present themselves, but possible remedies come to light as well. Two primary physical causes are inadequate bypassing of dc supplies and thermal hysteresis. Unfortunately, another cause can be differential delays in the application of feedback or predistortion correction to the input signal in a linearization system. The ramifications of (3.10) are truly widespread in the linearized PA business. It is therefore appropriate to look at some experimental measurements which further confirm that the theory is correct.

Figure 3.15 shows a test setup which can be used to measure AM-AM and AM-PM distortion dynamically in the envelope time domain. A two-carrier signal is generated which has a peak power level approximating to the 1-dB compression point of the test PA. Couplers sample the input and output signals which are fed into a suitable vector signal analyzer, which in this

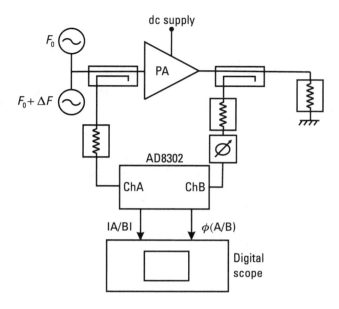

Figure 3.15 Test setup for dynamic AM-AM and AM-PM measurement.

case is an Analog Devices 8302 chip. This chip is a gain and phase detector which has two input RF channels, and gives independent differential gain and phase outputs which can be displayed on a digital oscilloscope. Figures 3.16 and 3.17 show a set of traces for a Class AB GaAs MESFET amplifier. The traces show a wide range of carrier separation frequencies, from 2 Hz up to 50 kHz. The 2-Hz gain and phase traces can be regarded as equivalent to a cw, or slow sweep measurement. It is immediately clear that as the separation, or modulation rate, increases, both gain and phase traces show substantial changes. Additionally, some hysteresis becomes evident in the 5–20-kHz range. This shows up as asymmetry in the corresponding spectral plots, shown in Figure 3.18; the low-hysteresis 2-kHz spacing case is shown for comparison.

Although the asymmetry is the focus here, these measurements show substantial overall changes in the gain and phase characteristics at higher modulation rates. This is an item of major impact on the nonlinear modeling of RF power transistors. It is clear, for example, that models which seek to curve-fit the static AM-AM and AM-PM characteristics will be fundamentally and possibly fatally inaccurate when applied to a dynamic signal environment. On the more positive side, it seems that once the lower kilohertz range has been passed, the curves settle down to a modified, but stable,

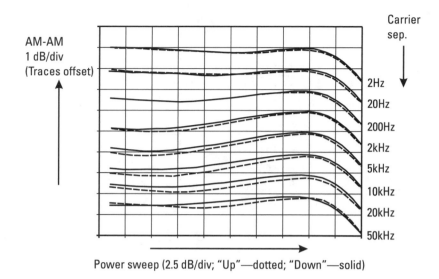

Figure 3.16 Dynamic AM-AM PA distortion plots.

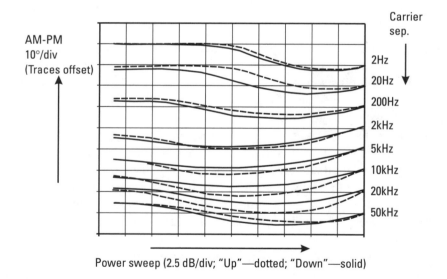

Figure 3.17 Dynamic AM-PM PA distortion plots.

appearance. A dynamic test of this kind would appear to bring new and valuable information to a modeling effort, although the general conclusions may vary from one device technology to another.

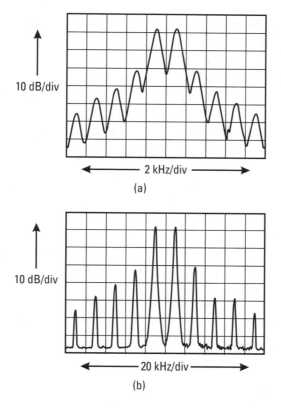

Figure 3.18 Corresponding spectrum plots for two-carrier dynamic envelope measurements (Figures 3.15 and 3.16): (a) 2-kHz carrier spacing and (b) 20-kHz carrier spacing.

The hysteresis, which seems to peak in the region of 10 kHz, would appear to be of thermal origin.[4] A GaAs die of the size used for these measurements will have a thermal time constant of the order of 100 microseconds or so. In analyzing the dynamic thermal characteristics of an RF transistor die, there are in fact three time regimes to be considered. The slowest of these, which can be of the order of a second, is the large thermal mass of the heatsink to which the die is attached. The second is the aforementioned thermal time constant of the bulk die, and the third is the active channel within the die which can show a time constant of a few microseconds. The

4. The dc supply was carefully monitored during these tests, and displayed no detectable modulation.

measurements here appear to show a thermal "resonance" in the second regime. It remains speculative as to whether another thermal hysteresis region may exist at megahertz modulation rates.

There is another cause of asymmetrical IM effects in RF power amplifiers. This is the very common problem of inadequate supply rail decoupling, and is illustrated in Figure 3.19. A deep Class AB PA draws current from the supply in synchronism with the AM of the input signal; the "dc" supply is now a high-level broadband video signal, possibly stretching up into the megahertz region for multicarrier signals. In high-power amplifiers, the current peaks can be several amps, and a large reservoir capacitor has to be provided physically close to the transistor in order to supply these peaks. Provided this capacitor has no parasitic resonances within the entire video bandwidth of the signal, and it has a large enough capacitance to supply the current peaks with negligible voltage drop, no adverse effects will take place. Figure 3.19 shows a typical case of inadequate supply decoupling. A substantial voltage modulation appears on the supply rail which will cause additional AM to be created. If the supply impedance is significantly reactive, this spurious AM will be out of phase with both the AM and PM of the signal. This

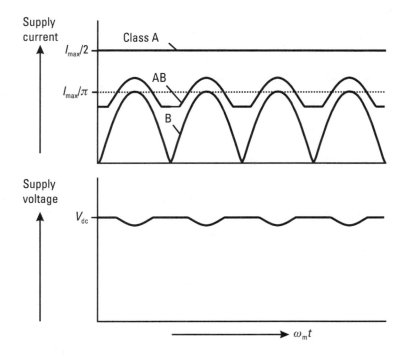

Figure 3.19 Supply rail modulation effect.

causes an envelope domain phase offset between the AM and PM emerging from the amplifier, and IM asymmetry will result.

Unfortunately, for modulation bandwidths which extend into the megahertz region, it becomes increasingly difficult to design a bias network which eliminates this effect. There is also another constraint on the placement of large capacitors in RF transistor bias networks, the increasing likelihood of video band oscillation. Most RF transistors will be intolerant of short-circuit terminations at frequencies in the megahertz through to the VHF range. Fortunately drain supplies of FETs are much less sensitive in this respect, although care still must be exercised. Bipolar transistors present a more formidable problem in megahertz data rate modulation systems. In some cases it may be necessary to accept that some spurious modulation will occur, and the bias network has to be designed to appear resistive. It has been shown that tuning of the bias network in the video band can satisfactorily eliminate asymmetrical ACP effects [6].

There is another possible cause of IM asymmetry, which lends itself to closer experimental characterization and PA design scrutiny. It has already been noted, in connection with (3.10), that elimination, or even reduction, of AM-PM effects will inevitably reduce asymmetry. It is therefore reasonable to ask where AM-PM effects come from in the first place, and whether a PA can be designed to reduce or minimize AM-PM effects. The answer to this seems not nearly so well-defined as the more obvious physical amplitude limiting process which causes AM-AM distortion. For RF power transistors, a primary cause of AM-PM effects appears to be the dynamic mistuning of the input match. This is illustrated in Figure 3.20, which shows the dependency of the phase of the small signal transmission parameter s_{21}, on the gate bias of a GaAs MESFET power device in both matched and unmatched conditions. The bias-dependent small signal measurement clearly reveals a substantial voltage dependency, and consequently a potential nonlinear effect under large signal conditions. The phase change is mainly caused by the C-V characteristic of the gate-source junction. In a practical amplifier design, this capacitance forms part of a resonant matching network, which although boosting the gain to useable levels, also greatly multiplies the phase-capacitance dependency. Unfortunately, the effect is also clearly magnified by operating a device in Class AB, where the quiescent operating point is much closer to the region of maximum phase variation.

A key issue with this particular nonlinearity is that it is associated with a high Q resonator and will be subject to delay, or latency, due to the build-up time of the resonance. This effect is considered in more detail in Chapter 4 (Section 4.5), but in this context it clearly represents a primary cause not

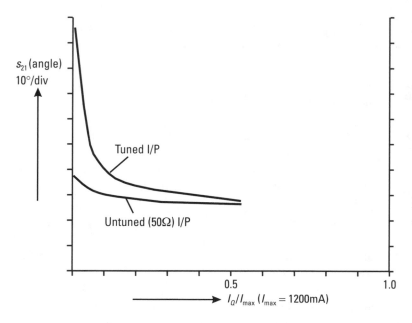

Figure 3.20 Bias dependency of s_{21} phase angle for GaAs MESFET power transistor.

only of AM-PM but also a time offset between AM-PM effects and output compression limiting. The two curves in Figure 3.20 represent the extremes of a possible design tradeoff in reducing AM-PM and asymmetry effects. It would appear that some deliberate mistuning on the high Q factor input match of large RF power transistors may pay off in terms of improved AM-PM performance for the loss of a decibel or two of gain. Better still, if a new generation of high-voltage RF power devices becomes available, giving higher power per unit cell periphery, major linearity benefits may become apparent due to the lower Q input matching requirements. Such technologies as Silicon Carbide and Gallium Nitride should be carefully evaluated even if they do not in the first instance appear to offer advantages over conventional devices in terms of raw power and efficiency specifications. In practice, the same observations and comments may be applicable to other parasitic capacitances which feature in most RF power devices. It is, however, noteworthy that the device used in the measurements shown in Figure 3.20 is a GaAs MESFET which displays quite low-voltage dependency on capacitive parasitics other than the input C_{gs}.

The observations and analysis in this section point clearly to an envelope time domain delay, or phase offset, between the AM-AM and AM-PM processes as being a root cause of asymmetry in IM and ACP spectra. Asymmetrical effects are troublesome, especially if predistortion is being used as a linearization technique (see Chapter 5). If the process can indeed be modeled as a simple delay between the two primary distortion processes, it would seem possible to extend the envelope simulation concept to include it. Instead of representing the output of a PA as being a function of the input envelope amplitude alone,

$$v_o(\tau) = G\{A(\tau)\}\cos(\omega t + \Phi\{A(\tau)\})$$

where $A(\tau)$ is the input envelope amplitude and τ is envelope domain time, the basic expression can be modified to

$$v_o(\tau) = G\{A(\tau)\}\cos(\omega t + \Phi\{A(\tau - T)\}) \tag{3.11}$$

This is probably an optimistic simplification of a highly complex physical process which ultimately defies analytical definition, but like many such approximations, it may be much better than nothing at all. The next logical step would be to recognize that both AM and AM-PM processes at the output are functionally connected not just to two specific times in the envelope domain, but to a convolution of a whole period of previous times. This represents a major increase in model complexity but has been reported [7]. The actual requirements for parametric characterization in this extension are quite formidable. In the end, there is a more viable pragmatic solution, which is to understand the mechanisms which cause memory effects and design the hardware such that the effects are minimized. This eliminates the need for a detailed simulation and modeling procedure, and the simple approximation of (3.11) may be adequate.

3.6 PAs and Peak-to-Average Ratios

There is probably no more intensely discussed topic in RFPA circles than the impact of high peak-to-average ratio signals. Multicarrier signals, and now even single-carrier signals such as WCDMA, can have peak-to-average ratios extending up to 10 dB and even beyond. This means, for example, that in order to transmit a 10-W signal, the PA has to be able to pass a 100-W peak, now and again. Here lies the crucial issue: If it only happens very infrequently, can we use a 20-W PA, blink at the right time and get away with it?

This may be viewed as a trivialization of a serious problem—a problem which has a major impact on system design and economics, and a problem which surely deserves a more serious quantitative characterization.

Yet in some respects, the problem thus posed would appear to answer itself. The peaks are there for a reason; they are part of the process of carrying information. It may be reasonably surmised that clipping a peak means clipping the information transmission process; there *will* be an effect and it will be in the form of a reduction in the information that was originally sent. So if information integrity is to be completely preserved, the peaks cannot be clipped. Curiously, even this intuitive argument is flawed. The reason for this lies in some very clever innovative work done several decades ago, which proposed some methods of conveying information on an RF carrier which could survive a moderate degree of peak clipping without affecting the integrity of the information transmission. For example, Figure 3.21 shows the phaseplane trajectory of a North American Digital Cellular (NADC) signal. This is a $\pi/4$ differential quadrature phase shift keyed (DQPSK) format, with a Nyquist raised root cosine (RRC) filter. The filter is a key element; it greatly reduces the bandwidth of the raw phase-switched RF signal, yet it does so in a manner which does not affect the information content or the fidelity of the demodulation process. This favorable situation inevitably comes at a price[5], and this takes the form of AM. The information in such a signal is carried entirely in the phase at the stipulated constellation points, also shown in Figure 3.21. The AM carries no information; it is overhead which has to be paid as a result of the greatly compressed signal bandwidth.

The interesting feature of the largest AM peaks, however, is that they occur *between the constellation points*. It is therefore possible to clip these peaks, to a limited degree, without causing any degradation in the bit error rate (BER). There will, however, be spectral distortion, and depending on the local laws and regulations, this may still limit the extent to which the PAs can be allowed to clip or saturate. This is an intended feature of this modulation system, and indeed for many related QPSK systems. It appears that there was a time, quite long ago, when the efficiency and cost of a system PA were considered to be of such paramount importance that modulation techniques were specially developed to minimize, or optimize, their imperfections. The modern wireless communications era seems to have emphatically bucked this trend, and places increasing demands on PA linearity as information throughputs are increased in narrow, fixed-band allocations.

---

5. This assertion is based on the principle of "Conservation of Grief," a valuable and usually quite reliable guideline in engineering problems.

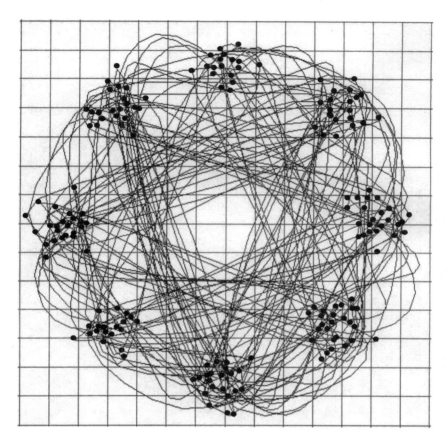

Figure 3.21 NADC signal constellation.

An example of the modern, unsympathetic approach to the PA requirements is illustrated in Figure 3.22. This shows a signal magnitude and phase
trajectory for an Enhanced Data Rate for GSM Evolution (EDGE) signal.
This is a much more efficient system, in terms of data throughput per unit
bandwidth, than NADC, even at comparable clock rates (which the two systems in practice do not have). But at first sight, the peak-to-average ratio
appears (and in actuality is) about the same as the NADC signal. The key
point about the EDGE signal is that some constellation points lie right at the
edge of the amplitude plot. So if the peaks are clipped, there will be an almost
immediate degradation in BER, as well as spectral spreading. The concept
of error vector magnitude (EVM) was defined and discussed in Section 3.3.
Modern system regulations (e.g., EDGE) attempt to quantify the degree of
allowable PA distortion by having an EVM specification. This has largely

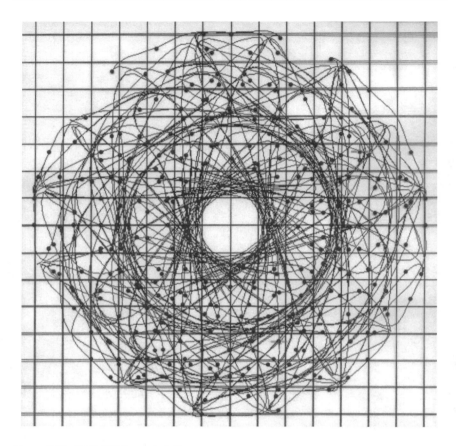

Figure 3.22 GSM EDGE constellation.

arisen out of the availability of suitable measuring equipment which can measure the EVM directly. Unfortunately, the regulatory embracement of the EVM concept has shown up some hazards. It seems that system manufacturers, and even writers of system standards, have quickly absorbed the concept of EVM specs because it provides a way of building margin into a system at the expense of the PA designers' sanity. There is obviously a correlation between EVM and BER for a given system, but the BER may be affected by many other things additional to the PA. In this sense, an EVM spec makes sense because it is a measurable and specifiable parameter for a PA. On the other hand, the EVM spec needs to bear a fair and reasonable correspondence to the actual BER and ACP specifications of the system. This applies in particular to QPSK systems which can, and were originally conceived, to tolerate some interconstellation point clipping.

The above reasoning becomes even muddier when considering multi-carrier signals. Multicarrier signals can generate very high peaks; the peak-to-average power ratio is ideally proportional to the number of individual carriers. It is much less clear, however, and much harder to quantify the impact of clipping these peaks. There are simply too many missing pieces of information to be able to give the much-sought generalized answer to this question. It is also a difficult problem to simulate. The best advice that a PA designer can take is to perform careful measurements, maybe at scaled-down power levels, in order to determine the extent to which the peaks can be clipped and all conceivable requirements safely met. Different system specs, different national emission standards, even varying levels of system usage, all conspire to make general analysis of the basic problem very challenging. Once again, it seems that an analog computer, consisting of a representative signal generator, the test device, a spectrum, and a vector signal analyzer, still gives us the answer we seek in a second or so. This is, for the time being, a lot faster than either digital simulation or mathematical analysis.

For all the debate and discussion which takes place on this subject, not to mention the volumes of analysis appearing in the literature, the "answer," in any individual case, is a simple number, expressed in decibels. This number represents the acceptable shortfall in PA capacity from the overall PEP level, which enables specifications to be met. It is often greater than zero, but rarely as great as the overall peak-to-average power ratio. This is not, in many cases, such a large range of uncertainty. It is even relevant to speculate that part of the reason for the protracted debates on this subject is the usually absent definition of what, precisely, is meant by "average" power. For many RF engineers, average power is the reading observed on a thermal power meter. Such readings, and indeed such instruments, have a rather limited relevance in a high-speed amplitude modulated RF system. In such an environment, the measurement and specification of average power give an unfairly biased view of the power requirements of the system, as viewed in a more relevant time domain, such as the envelope domain. So, for that matter, does a spectrum analyzer whose trace is averaged over similar time periods, yet this seems to be common practice.

3.7 Conclusions

The characterization and effective modeling of RFPAs is still a topic of worldwide research. Complex broadband modulation schemes and memory effects in high-power PAs compound an already challenging problem. The tools and modeling techniques which are required to perform a complete and accurate simulation of a high-power PA in a modern multicarrier communications environment are not yet available. But a less ambitious approach to modeling and simulating PAs with simpler models and modulation systems can be very instructive. The intuitive but approximate simplification to the simulation problem offered by so-called envelope domain techniques provide a useful tool which can give acceptable results in many cases. PA models for such simulators can be derived using two-carrier dynamic, rather than single-carrier static, measurement data. Such a change in modeling basis will in itself reduce the impact of PA memory and hysteresis effects on the simulation process.

The difficulties of simulation in some respects belie a surprisingly straightforward practical situation. Techniques exist for preserving mean power consumption of RFPAs which have high peak power capability. Simple benchmark testing, using representative devices and test signals, can provide direct and speedy answers to the peak power requirements of a given modulation environment.

References

[1]   Kenney, J. S., and A. Leke, "Simulation of Spectral Regrowth, Adjacent Power, and Error Vector Magnitude in Digital Cellular and PCS Amplifiers," *Microwave Journal*, October 1995.

[2]   Volterra, V., *Theory of Functionals and of Integral and Integro-Differential Equations*, New York: Dover, 1959.

[3]   Maas, S., *Nonlinear Microwave Circuits*, Norwood, MA: Artech House, 1988.

[4]   Saleh, A. M., "Frequency-Independent and Frequency-Dependent Nonlinear Models of TWT Amplifiers," *IEEE Trans. on Communications,* Vol. COM-29, No. 11, 1981, pp. 1715–1719.

[5]   Kaye, A. R., D. A. George, and M. J. Eric, "Analysis and Compensation of Bandpass Nonlinearities for Communications," *IEEE Trans. on Communications,* Vol. COM-20, No. 10, pp. 965–972.

[6]   Sevick, J. F., K. L. Burgher, and M. B. Steer, "A Novel Envelope Termination Load-Pull Method for ACPR Optimization of RF/Microwave Power Amplifiers," *Proc. Intl. Microw. Symp.,* MTT-S, Baltimore, MD, 1998, WE3C-7.

[7]   Ngoya, E., et al., "Accurate RF and Microwave System Level Modeling of Wide Band Nonlinear Circuits," *Proc. Intl. Microw. Symp.,* MTT-S, Boston, MA, 2000, TU2D-1.

4

Feedback Techniques

4.1 Introduction

Feedback techniques are the top, bottom, and sides of electronics as we know it. Since the earliest electronic era, negative feedback has been recognized as an indispensable ally in harnessing the amplifying capability of quirky electronic devices. As with most epoch-making inventions, the precise chronology of its origins is somewhat in question, and assertions of priority are usually tainted by nationalistic bias. It is clear that Nyquist, working at Bell Labs in the early 1930s, laid much of the foundations which still form the basis of feedback theory. But the actual term "negative feedback" was first used in the context of electronic amplification in a 1933 patent, by the prolific British electronics engineer Alan Blumlein [1, 2]. A 1929 patent by Black, at Western Electric [3] is frequently quoted as the first disclosure of a feedforward linearization scheme; curiously this predates Black's own feedback patent which appeared in the United States 4 months after the Blumlein patent, in January 1934.

The intention in this chapter is to treat the subject of feedback with the respect it deserves in a broader electronic context, despite the fact that the only dissenters seem to be those who design microwave amplifiers. They claim that feedback techniques, as conventionally employed at lower frequencies, come to irrevocable grief due to the much larger phase shifts and electrical lengths that amplifying devices display at gigahertz frequencies. But so-called "indirect" feedback techniques, which operate with the detected, or

111

downconverted, envelope amplitude and phase variations, offer some possibilities for bypassing the fundamental phase delay problems which certainly do seem to eliminate the use of conventional, or direct, feedback methods. Unfortunately, even the indirect RF feedback methods suffer from the same basic problem that causes instant dismissal of the conventional feedback approach in the first place; electrical delays around the feedback loop restrict the bandwidth of signals that can be linearized, and can still ultimately lead to instability. Such limitations have restricted the usefulness of these well-established techniques in modern multicarrier systems. Indirect methods have also suffered from another disadvantage in the present context; they are conventionally formulated as *transmitter*, rather than *PA* techniques, and rely fundamentally on access to a baseband, or IF, signal input. In fact, the Polar Loop is frequently mis-assigned under the indirect feedback category; it is really a signal reconstruction technique such as those discussed in Chapter 2.

In a modern wireless communications context, the PA designer appears to fall between two stools: Conventional direct feedback is out of consideration due to the multiple nanosecond group delays which are required for signals to transit higher power amplifiers; indirect feedback techniques become marginal as the instantaneous signal bandwidth, such as in multicarrier applications, increases to the point where even these few nanoseconds of delay can seriously limit and degrade the effectiveness of the technique. But there seems to be a narrow window of hope in this generally gloomy surmise. If indeed the delay can be reduced to literally a few nanoseconds, rather than tens of nanoseconds, indirect feedback methods can surely still be used with great effect and little overhead compared to the open loop techniques which currently dominate the MCPA industry.

This chapter reviews and calls into question some of the conventional wisdom on this subject. It represents, in some respects, a case for the defense for amplifier feedback techniques. Feedback methods enable the audio designer to achieve linearization performance some way in excess of that required in modern communications MCPAs, with the simplest of circuitry, without any need for gain and phase tracking, built-in test equipment, pilot tones, or DSP controllers. Much the same applies to all low-frequency analog design, to the extent that practitioners don't even need to bother learning about nonlinear effects in their devices; it's all knocked out of sight by the feedback. There are very good reasons why things are not so easy at gigahertz frequencies; however, the microwave linearization alternatives are cumbersome, inefficient, and less effective by comparison. Diligence, innovation, and DSP have all made these alternatives viable, but the situation remains frustrating. Any open loop technique which seeks to reduce distortion products by 30 dB or more has to include an overhead structure of sensing, monitoring, and correcting for the smallest changes in the physical and electrical environments, not to mention drift in the characteristic of the PA components, and even the monitoring system itself. Under limited conditions, it can all be made to work, but volume production of such systems remains a challenge which few have demonstrably met.

Against this background of adaptive open loop system design, with its cumbersome overheads, is the RFPA delay, or *latency*, problem indeed insurmountable? Given that a typical feedforward system, with all the overheads, gives a typical MCPA product with only just over 10% efficiency, are there hitherto unconsidered tradeoffs in the basic PA design which would reduce the inherent delays to a level which allows feedback to be used as the main linearization element? The answer, it seems, is maybe. A new generation of high bandgap RF power devices, coupled with well-established RFIC and hybrid circuit techniques, can together modify the rulebook on PA design tradeoffs, with a much stronger emphasis placed on the latency issue. The high bandgap devices can alleviate the matching Q-factor; the use of devices with relatively high cutoff frequencies can result in broader bandwidth designs having lower delay, and RFIC or hybrid techniques can greatly cut down on the length of interconnections. Feedback is such a good solution that it seems worthy of a little more effort to see whether there is any possibility at all that it may be harnessed at higher frequencies.

One of the primary stumbling blocks for the success of feedback techniques of any kind in microwave amplifiers is the need for high Q matching networks to match the very low impedances of high-power microwave transistors. The high Q-factors cause latency in the response time of the amplifier. The recent emergence of new PA device technologies, such as Silicon Carbide (SiC) and Gallium Nitride (GaN), having high-voltage operation and higher matching impedances may lead to a change in this situation, and re-open the door for feedback design. As a minimum, the design of microwave PAs with lower group delay may enable indirect feedback methods to work in multicarrier applications. Here we will be concerned with such indirect feedback methods as can be employed around the PA itself, without conversion to, or need for knowledge of, the baseband signal. In this area we will introduce the vector envelope feedback loop, as distinct from the reconstructive Polar Loop, or the downconverted Cartesian Loop, both of which are sidelined here as essentially transmitter, rather than amplifier, techniques. Detailed analysis of the Cartesian and Polar Loops, as conventionally defined, is outside the scope of a book on RFPA design techniques. The reader is referred to an excellent in-depth treatment [4] and a substantial literature for further reading on these subjects.

It is appropriate to summarize the goal in the ensuing sections of this chapter. We seek nothing less than a feedback PA linearization scheme that displays the same level of linearization as is typically reported using conventional feedforward methods. The conclusions will show that for the most demanding multicarrier signal environments this goal may not be achievable using current PA device technology and design methods, but realistic targets can be set for a new generation of PA designs. Less demanding signal environments will be shown to have significant possibilities for feedback linearization, and the frequently quoted limitations of this approach can be reduced by careful design of the various RF components.

4.2 Feedback Techniques

This section gives a brief review of the basic feedback methods available to the RFPA designer. Some more detailed analysis, especially of some envelope feedback methods, will be given in later sections. The goal in this section is mainly to define terminology and review the alleged limitations of the various techniques.

Direct feedback (Figure 4.1) should need no introduction. As stated above, it is truly the cornerstone of modern electronic design. Throughout the electronic age, from the earliest vacuum tubes to the modern integrated circuit, the problem of maintaining consistent performance from one device to the next has been almost entirely dependent on feedback, and often lots of it. The elementary relationships showing the reduced dependency of the gain of a feedback loop on the gain of its active element can be found in any textbook, and are reproduced in Figure 4.1. Analysis shows that providing the inverse phase connection of the feedback signal is maintained at the input, the linearity of the overall loop is improved as an inverse function of the gain reduction, as compared to the active device running open loop. In RF terms, this means that a 10-dB reduction in gain, due to the use of a feedback loop, results in a 10-dB lowering of any nonlinear product caused by the variations in the device gain characteristic. Thus an audio designer can confidently take an open loop gain block having 100 dB of gain, and put 50 dB of feedback around it, to make a "hi-fi" amplifier at a useful gain level.[1]

---

1. In fairness to the audiophile, really high-quality audio amplifiers are not quite this easy.

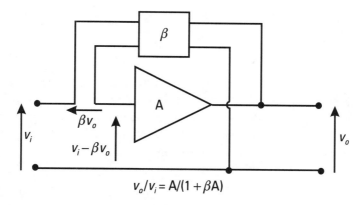

$$v_o/v_i = A/(1 + \beta A)$$

Figure 4.1 Classical feedback amplifier configuration.

Life is not so simple at RF, we think. In order to obtain a useful reduction in IM products, we certainly need to think in terms of a gain chain of maybe 60 dB, in order perhaps to use 30 dB of feedback. This would reduce IM or ACP products by 30 dB, without affecting the overall efficiency significantly. A PA having 100-W output and 30-dB overall RF gain requires a drive signal of 20 dBm, which is obtainable from a cheap RFIC driver. Such a device would, of course, be needed in the first half of the gain chain itself, to make up the extra 30 dB of open loop gain. But the key point is that the extra gain is cheaply obtained, even at 2 GHz using modern RFIC components. Clearly, such an amplifier would represent a dazzling revolution in RFPA design and performance. So where is the revolution?

The answer is the stumbling block of signal delay. A multistage PA having 60-dB open loop gain will typically have such a rapid in-band phase versus frequency variation that the 180° phasing of the feedback signal can only be maintained over a very narrow frequency band. Indeed, in a conventional PA design which has to hold a specified performance over a substantial RF bandwidth, the loop may well oscillate at an in-band frequency. Oscillation is frequently cited as the main hazard, but in fact such an extreme possibility can usually be suppressed by careful RF design. A more restrictive problem lies in the fact that the calculated benefits of negative feedback only apply at the 180° point, and erodes away either side as the phase response rotates away from the ideal value. A full derivation of this can be found in many electronics textbooks, and is only summarized here; if the gain A is replaced by a complex gain, $Ae^{j\phi}$, and the overall gain amplitude $v_o/v_i = G$, the feedback gain equation can be written as

$$G = \left| \frac{Ae^{j\phi}}{1 - \beta Ae^{j\phi}} \right|$$

which can be differentiated to give an expression relating the variations in the amplifier gain magnitude A in terms of the corresponding variations in overall gain G. The result, after some manipulation and making the customary approximations concerning the magnitudes of the feedback factor β and the open loop gain A, can be written in the form

$$\frac{\partial G}{G} = \frac{1}{\beta A} \frac{\partial A}{A} 2\cos\phi$$

which is plotted in Figure 4.2, where the basic 180° phase shift has been assumed, and the horizontal axis represents phase deviations from the ideal result. So in order to obtain the full classical feedback linearity benefits, the open loop phase response can only stray from the 180° point by a matter of a few degrees, although it is significant that the degradation, measured in decibels, does not escalate until the 45° point is passed.

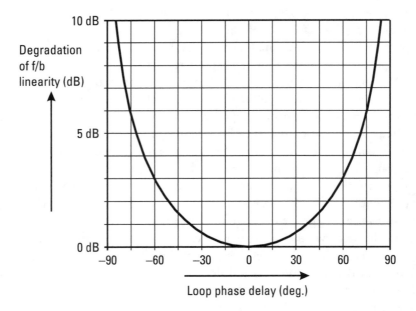

Figure 4.2 Degradation of feedback linearization effect due to phase shift around the loop.

A measurement of a typical 2-GHz HPA may show a phase delay of as much as 20 nsec. This represents a 45° bandwidth of just over 6 MHz.[2] Oscillation will become a certainty within a 25-MHz bandwidth, with the assumption that this will almost certainly lie within the amplitude passband.[3] These numbers effectively represent the final condemnation of the concept as far as microwave (> 1 GHz) applications are concerned. There is a final catch in that this loop bandwidth has to include the frequency span occupied not just by the original signal, but the ACP or IM products which the loop seeks to suppress. Yet there are possibilities for significantly reducing the void of practicality that these numbers represent. These possibilities form a focus for this chapter, since we will see that they also open up new possibilities for indirect feedback also.

The block diagram of an envelope feedback scheme is shown in Figure 4.3. The basic concept is simple and appealing, and has been used by transmitter designers from the earliest days of wireless communication. Such schemes will, in this book, be described as "envelope feedback" techniques, since they seek to force the detected characteristics of the RF envelope, either amplitude alone or amplitude and phase, to be equal at input and output. These methods are *bona fide* feedback techniques, and should be always regarded as such; they are by no means just poor relations of direct feedback. What they recognize is that in a typical band-limited RF communications

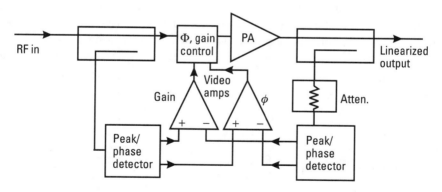

Figure 4.3 Envelope feedback system.

2. Assuming a definition of 45° bandwidth as the frequency shift required to obtain a 45° phase shift.

3. To make matters worse, further attempts to restrict the bandwidth by conventional reactive filtering will almost certainly result in a higher group delay and a narrower stable bandwidth.

system, the harmonic distortion products of the RF carrier can be removed by filtering, and so it is only necessary to remove or reduce the harmonics of the modulation signal which are generated by nonlinearities in the RFPA. Provided both amplitude and phase correction are included, such methods are capable of giving the same linearization benefits as direct feedback, with the important advantage that the loop gain is mainly provided at baseband, rather than at the carrier frequency.

As with direct feedback, it is not difficult to find reasons why such schemes may not work at higher RF carrier frequencies. This is particularly the case for the simplest form of envelope feedback, which has only an amplitude correction loop (Figure 4.4). Such a system can actually be viewed as an envelope predistorter (see Chapter 5) with a "smart" analog driver. As discussed in Chapter 5, one problem with such a device is that as the RFPA distorts, the required amplitude correction at the input escalates as the compression region is reached. This means that the AM-PM will be degrading simultaneously as the input amplitude controller strives to correct the gain compression. There can be an additional related problem in that the amplitude control device will almost certainly introduce some AM-PM of its own. In the more backed-off drive regime, higher video gain may be required in order to maintain useful linearization action. All of these issues can, in principle, be handled by careful design of the various components, and the inclusion of a phase correction loop. The fundamental limitation of an envelope feedback system is just the same as in the direct feedback case: phase delay round the control loop. The same calculation can be performed for a typical 20-nsec delay, giving the same estimate in terms of the signal bandwidth (6 MHz) for which the system can remain near to optimum

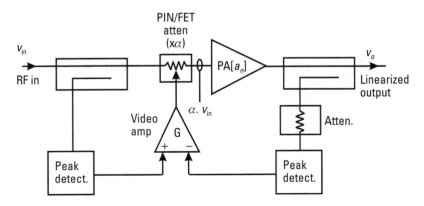

Figure 4.4 Basic amplitude envelope feedback system.

linearization performance. It should be fairly noted, however, that in the envelope feedback loop the delay includes detection and video amplification processes.

The advantages of including an envelope phase feedback loop was recognized many years ago, and gave rise to two widely used transmitter architectures, the Cartesian and Polar Loops (Figures 4.5 and 4.6). Both methods implement phase and amplitude envelope feedback, but to achieve this they convert the PA output to baseband (or IF) and perform the necessary input-output comparisons and error amplification at the lower frequency. The Cartesian Loop is a straight vector envelope correction scheme which forces the amplitude and phase of the input and output signal envelopes to be the same,

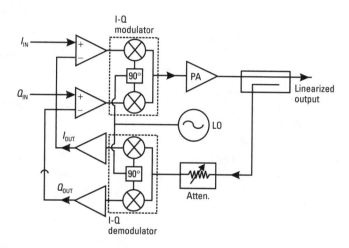

Figure 4.5 Basic Cartesian Loop linearization system.

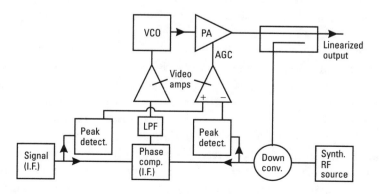

Figure 4.6 Polar Loop system.

the residual error being a function of the loop gain and the linearity of the downconverter. The Polar Loop, as has already been commented, is a radically different architecture in the sense that it constructs the complete RF signal starting with a cw source. This source is given the appropriate phase modulation using a phase-locked loop and then the AM is applied. The action of control loops forces the output envelope to track the input signal in both amplitude and phase. Such a scheme should really be classified as a derivative, or superset, of the Khan restoration loop (EER), in that it constructs a linearized signal, rather than linearizing an existing one. Neither of these schemes is applicable to an RFPA, which only accepts a modulated RF signal, either single or multicarrier.

It is, however, quite possible to conceive of methods by which phase and amplitude envelope feedback can be applied around an RFPA. Such methods will here be termed "vector envelope feedback." These methods will be covered in a later section, but should be considered as quite distinct from the "Polar Loop" as conventionally described in the literature. One of the disadvantages of the direct, or vector, feedback loop is that the envelope detection process generates video signals which have higher bandwidth than the original modulation bandwidth of the RF signal. This can be seen very simply in the envelope detection of a two-carrier RF signal; the detected signal has the appearance of a full-wave rectified version of the RF signal envelope and contains a string of even harmonics of the original modulating signal; such detectors are, fundamentally, frequency doublers. Amongst the various problems associated with this form of feedback, which will be discussed in subsequent sections, this would appear to be the least obvious and possibly the most restrictive. Against this fundamental disadvantage, however, can be set the simplicity and compactness of the RF detectors.

The basic statement that high-frequency devices, operating at gigahertz frequencies, have high signal delays seems at first sight contradictory. In order to amplify a microwave signal, the delay associated with the transconductive element in a microwave transistor has to be measured in terms of a few picoseconds, and indeed this is usually the case, perhaps stretching to tens of picoseconds in some cases. Yet the group delay, even of a single high-power 2-GHz stage using such a device, can typically be measured in the range of several, or even over 10, nanoseconds. This is essentially due to the high Q-factor associated with the matching networks, but this basic cause can be substantially multiplied through the inadvertent, but common, use of regenerative effects in the quest for maximum gain from higher power microwave gain stages. The net result, accumulated over several gain stages and coupled with convenient but inappropriately spacious layouts, can be a high-power PA with a delay of tens of nanoseconds. Such an amplifier will have such rapid phase rotation as a function of frequency that direct negative feedback (Figure 4.1) becomes a contradiction in terms, oscillation being a possibility even in-band. The impact on an indirect feedback scheme, such as that shown in Figure 4.3, will be less drastic but highly restrictive neverthe- less. The closed-loop delay time in effect sets an upper limit to the rate of change of input signal that can be effectively linearized, and also sets a limit on the allowable loop gain from stability considerations.

It is a curious observation that although the detailed mathematical analysis of the feedback systems shown in Figures 4.1 and 4.2 is well repre- sented in the literature, the main independent variable in any such analy- sis—the amplifier delay—is treated as an indispensable burden about which little can be done. In fact, even if this delay could be reduced by a factor of 2 in any given situation, an important benefit would be obtained. Although the exact analysis is complicated, such reduction in the amplifier delay will surely reduce the critical rate of change of phase versus frequency, $d\phi/d\omega$ by the same factor, and it is a fair assumption that the signal bandwidth limit could be expected to improve by a similar ratio. So here is a strange irony; microwave amplifier designers pay no attention to the delays in their designs because they are not regarded as a critical parameter. System level designers typically purchase amplifiers as finished items and seem to accept high-delay parameters. The focus in Section 4.5 is to explore the causes of amplifier delays and to quantify how much reduction may be possible. If an order of magnitude reduction is possible, reducing the PA delay into the nanosecond regime, then direct feedback techniques may have much wider applications than currently assumed.

4.3 Amplitude Envelope Feedback: Configuration and Analysis

It is appropriate at this stage to consider the basic amplitude envelope feed- back loop in more quantitative detail. There are still applications where the PA latency issue is not a fundamental limitation to using the technique, and some consideration needs to be given to the realization of the other compo- nents in the loop in order that useful linearization can be obtained. The very desirable inclusion of an envelope phase correction will be considered in Sec- tion 4.4. This potentially valuable addition to an RFPA seems to have been curiously overlooked; the assumption seems to be made that in order to include phase correction it is necessary to make the major step up to a system level implementation such as the Cartesian or Polar Loop, where the phase comparison is performed at an intermediate frequency (IF), or baseband, frequency. In principle phase detection is quite feasible at RF without down-conversion. The availability of higher quality RF components and RFICs up to 2 GHz appears to make this option worthy of further investigation.

Figure 4.7 shows an RFPA with a basic amplitude envelope correction loop. The input and output detectors are assumed to be perfectly matched, and the associated coupling and attenuation values are chosen such that the linear gain of the PA is cancelled, so that the detectors receive identical RF signal levels in linear operation. The detectors are fed into a "video" differential amplifier, generating an error signal which drives an input attenuator in the appropriate sense to restore the error sensed by the detectors. Even in this qualitative description of the feedback system, there is an immediate flag of caution to be raised. The attenuator drive characteristic has to be carefully chosen for correct operation of the system. It is not simply a matter of getting the drive to change the PA input amplitude in the appropriate direction—that is, for example, to reduce the attenuation as the amplifier is driven into compression. When the goal is to linearize an otherwise nonlinear component, it is important to ensure that the system still operates in a useful manner, even when the amplifier is operating in its linear region. Basically, this problem can be traced to the output voltage at the differential amplifier. In ideal conditions, this will drop to zero when the PA drive is backed off into the linear regime. The attenuator characteristic must therefore be chosen to have a "sensible" median value as its drive voltage crosses zero. Although this may seem an obvious requirement to an analog designer familiar with automatic gain control (AGC) control techniques, it is something of a trap for the RF engineer, who may well wish to evaluate a prototype system using

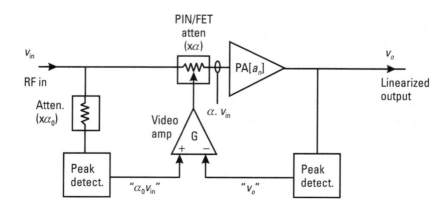

Figure 4.7 Schematic for analysis of envelope feedback system.

standard coaxial laboratory components for the detector and attenuator functions. Such components, frequently discovered lurking in a laboratory drawer or an absent colleague's toolbox, are typically unipolar in operation. A successful implementation of any envelope feedback scheme will usually require custom-designed RF components.

It was stated in the introduction to this chapter that envelope feedback of this kind can be categorized as a form of predistortion. In effect, the input attenuator is predistorting the signal envelope just as a conventional pre-distorter with a DSP driver does (see Chapter 5), except that in the present system, the drive is derived using "fed-back" information. The system is, however, subject to some of the same limitations as a predistorter, considered in detail in Chapter 5. Two of these limitations are worth previewing. First, the concept of increasing the drive level into the PA in order to restore the output to its linear level runs into a fundamental limitation as the amplifier compresses, and there is a well defined "point of no return," where no further increase of input drive can increase the output to the required level. This will, in practice, limit this system to somewhere in the region of the PA 1-dB compression point for useful operation. This can be a severe limitation when dealing with signals having a high peak-to-average ratio.

The second limitation is the bandwidth of the feedback loop. It is shown in Chapter 5 that the signal entering the PA, following the predistortion action of the attenuator, now contains spectral components which correspond to the higher-degree nonlinearities which the process seeks to remove. The predistorted signal emerging from the gain control element will display spectral distortion very similar to the uncorrected PA at the same drive level. Since the detection process itself can be considered to be a video frequency doubling function, and then if distortion up to the fifth degree is to be effectively corrected, the PA input signal must bear appropriate levels of fifth-degree components. It is therefore a quite realistic estimate that the components in the detection and video loop require a video bandwidth which is at least an order of magnitude higher than that of the original, undistorted input signal.[4]

Taking all these limitations as being met, and including the additional idealizing assumption that the input attenuator introduces no phase modulation as it varies the signal amplitude, it is instructive to analyze the system using a simple third-order nonlinearity for the PA. This will enable some

4. This explains, in large part, why the output spectra of such linearized PAs often show degraded higher order IM products; seventh- and ninth-order IMs can fall outside the video bandwidth of the correction loop and actually show degraded levels.

quantitative design information to be obtained about the relationship between the video gain and the degree of linearization available. The analysis initially assumes a quasi-static response to signal envelope variations; this is equivalent to assuming the delays of the PA and the linearization loop are negligible.

Referring to Figure 4.7, this analysis takes place entirely in the envelope domain, so that the signal voltage at any point in the chain, $v(\tau)$, represents the amplitude of a sinusoidal RF carrier varying in the envelope, or modulation time domain. In keeping with a convention used in this book, envelope time is denoted by (τ), as a reminder that envelope time is assumed to be several orders of magnitude slower than the variations of the RF carrier itself. The detector response is, for convenience, assumed to be ideal peak detection, so that the RF signal voltage amplitude appearing at the amplifier RF output is sensed by the detector as an equal voltage in the envelope domain. Clearly, the peak detectors do not respond to RF phase changes.

Note that the attenuator characteristic has been chosen such that at zero drive voltage there is a median value of attenuation, represented by α_0. The value of α_0 must be selected to allow for sufficient variation either side of this value to absorb any gain compression or expansion over the projected operating range.

The two basic equations describing the system are, first, the third-order PA characteristic,

$$v_o = a_1\left(\alpha v_{in}\right) - a_3\left(\alpha v_{in}\right)^3 \qquad (4.1)$$

where α is the attenuator setting, expressed as a voltage ratio. Therefore, α is itself a function of the output voltage from the differential video amplifier, following the law

$$\alpha = \alpha_o + G\left(\alpha_o v_{in} - v_o\right) \qquad (4.2)$$

where G is a composite video gain which is the product of the differential voltage gain of the video amplifier, and the attenuation drive characteristic measured in units of voltage attenuation per envelope voltage unit. (The input detector has a fixed attenuator of α_0 in order to have zero differential output in the median condition.)

The goal in the ensuing analysis is to solve (4.1) and (4.2) to obtain an expression for the composite linearized PA response, in the form

$$v_o = f\left\{v_i, a_1, a_3, G, \alpha_o\right\}$$

This is most conveniently obtained by solving initially for the attenuation α, as a function of input signal level v_{in}. Substituting for v_o from (4.2) into (4.1) gives, after some algebraic manipulation, a cubic equation for α,

$$\alpha^3 - \alpha\left[\frac{Gv_{in}+1}{a_3 Gv_{in}{}^3}\right] - \frac{\alpha_o\left(1-Gv_{in}\right)}{a_3 Gv_{in}{}^3} = 0 \tag{4.3}$$

The use of a simple third-order PA characteristic results in a cubic having three real roots (the so-called irreducible cubic) up to the point of saturation where two of the roots become imaginary. Some care is therefore required in using an iterative routine to find the single relevant root. But the solutions to (4.3) do give a functional relationship between the dynamic setting of the input attenuator, α, the composite video gain selection G, and the basic non-linear characteristic of the PA defined in this simple case by a_3.

Figure 4.8 shows the resulting gain characteristics obtained by solving (4.3) for several values of G and for a 20-dB input power sweep which takes the uncompensated PA up to just beyond its 1-dB compression point. Qualitatively, this plot shows a sharp reduction in the gain compression but the "point of no return" is evident, corresponding to a collapse in the linearizing

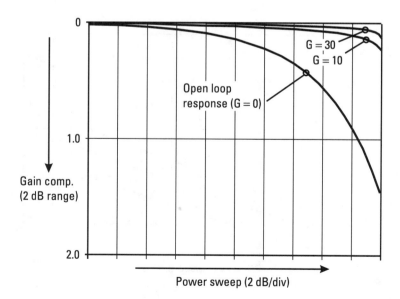

Figure 4.8 Quasi-static response of amplitude envelope feedback loop, for different video gain settings (ideal third-degree PA distortion).

action of the system. A more detailed quantitative picture of the same results is shown in Figure 4.9, where the drive has been turned into a two-carrier signal by applying sinusoidal modulation to the envelope. The third-order IM plots show three regimes: the middle regime, where the IM products are reduced by approximately the same factor as the video gain selection, the upper regime corresponding to the collapse of the linearization action due to saturation of the PA, and a third regime at very low drive levels where higher video gain is required to maintain the mid-regime IM correction. This third, small signal regime is where the main difference is seen between an envelope feedback system which uses gain control and a direct feedback configuration which controls the input level by subtraction of the feedback signal. As has already been demonstrated, the envelope or gain control system does not lend itself to the neat closed form analytical solutions obtained in a classical subtractive feedback system. Although Figure 4.9 appears to display the basic feedback "rule of thumb," whereby the reduction in distortion products is traded with amplifier gain, the terms of reference have shifted as well. In this case, the PA RF gain has not been compromised (other than 2–3 dB of median input attenuation), and the gain levy is extracted at the less sensitive video band. This is a very positive aspect of envelope feedback; on the debit side is the inescapable increase in delay time, which the use of higher video gain will incur. The other negative aspect, as already observed, is the

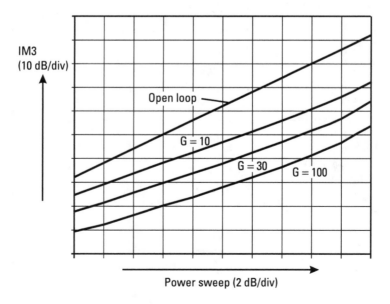

Figure 4.9 Two-carrier IM3 response of PA with amplitude envelope feedback.

reduction in IM correction at well backed-off drive levels. In practice, this may not be an important consideration, so long as the closed loop system still performs better than in the open loop state. The results of Figure 4.9 are well worth pondering, since they are a tantalizing illustration of how easy things would be in the linear RFPA world if PA and video process delays could be reduced to the negligible levels that these results assume.

The next step is to insert a realistic delay into the loop, as shown in Figure 4.10. Such an addition seems a relatively modest but necessary complication to restore the model to a real-world condition, but greatly increases the complexity of an analytical solution. By way of underlining this assessment, it is worth writing down the basic loop equations for the system shown in Figure 4.10. As before, a simple peak detection characteristic can be assumed for the detectors. The attenuator characteristic, at envelope domain time τ, can be expressed in the form

$$\alpha(\tau) = 1 + G\left\{ v_o \left(\tau - \Delta_{OUT} - \Delta_{VID} \right) - v_i \left(\tau - \Delta_{IN} - \Delta_{VID} \right) \right\} \quad (4.4)$$

where Δ_{IN} is the combined delay of the input detection and sensing circuitry.

The RFPA characteristic is defined to be

$$v_o(\tau) = \alpha \mathbf{a} v_i \left(\tau - \Delta_{PA} \right)$$

where $\mathbf{a}$ is the gain of the PA, written in bold type to indicate its underlying nonlinear content, and Δ_{PA} is the PA delay. So the overall input-output characteristic can be expressed as

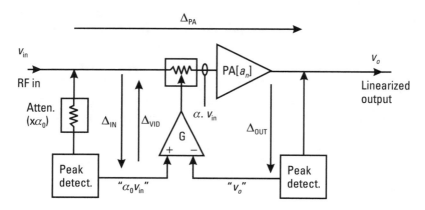

Figure 4.10 Delays in envelope feedback system.

$$v_o(\tau) = \left[1 + G\{v_o(\tau - \Delta_{OUT} - \Delta_{VID}) - v_i(\tau - \Delta_{IN} - \Delta_{VID})\}\right]av_i(\tau - \Delta_{PA}) \quad (4.5)$$

Equation (4.5) illustrates the analytical and simulation problems caused by the various delays, and is not proposed as a basis for further analytical treatment. Indeed, if the delay factors are even a rather small fraction of any significant envelope domain signal change, finding a solution appears to give indigestion to simulators, let alone theorists. As in *RFPA*, it is worth quoting a classical paper on RFPA linearization by Seidel [5], referring to the problems associated with delays around a feedback system:

> Feedback, a technique that has been used with much success, attempts a basic causal contradiction: *after an event has occurred, reshape its cause;* this violation may only be resolved by time smearing the event to blur the distinction between "before" and "after" to an adequate degree.

In another paper [6] the same author is more specific:

> Feedback, in comparing input with output, glosses over the fundamental distinction that input and output are not simultaneous events and, therefore, not truly capable of direct comparison. In practice, they are substantially simultaneous if device speed is far faster than the intelligence rate into the system … If we were to organize a system of error control which did not force a false requirement of simultaneity, not only would the problem of fabricating zero transit time devices disappear, but so would the entire problem of stability, another consequence of *comparing the incomparable.*

The author is, of course, leading up to a description of a feedforward system. In one respect, it is possible to take issue with this succinct but gloomy evaluation of RF feedback techniques. There is one simple addition to the envelope feedback system which offers some alleviation on the "comparison of the incomparable," by taking just one step towards a feedforward system architecture. A suitable delay can be placed on the input detector, so that both input and output detected signals arrive at the video differential amplifier at equal times in the envelope domain (Figure 4.11). In the terminology of Figure 4.10,

$$\Delta_{IN} = \Delta_{PA} + \Delta_{OUT}$$

This means that a "correct" error signal is generated, but at a delayed time as viewed at the input. The causality conflict in the application of the

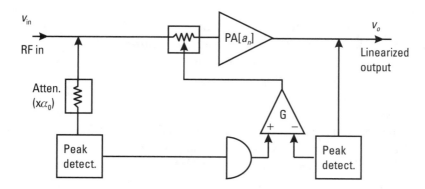

Figure 4.11 Use of input delay to correct time errors on formation of error signal.

correction therefore remains, but this addition seems, at least intuitively, to offer some alleviation.

Some quantitative reasoning further underlines the value of this important modification. Assuming a simple two-carrier input signal having an envelope amplitude function

$$v_i(t) = V \sin(\Omega t)$$

and a PA delay of Δ, where $\Delta = \phi t$, then the PA output envelope will be

$$v_o(t) = a_1 V \sin(\Omega t + \phi)$$

so that assuming ideal peak detection, the differential input to the video amplifier will be

$$\begin{aligned}v_o(t) - v_i(t) &= aV\{\sin(\Omega t + \phi) - \sin(\Omega t)\} \\ &= 2a\sin(\phi/2)\cos(\Omega t + \phi/2)\end{aligned} \tag{4.6}$$

So the "correction" signal, amplified by the differential amplifier, now contains a large corrupting sinusoidal component caused simply by the RF delay of the PA itself. Clearly, if the delay is small enough in comparison to a modulation "event," here conveniently assumed to be a single modulation period $2\pi/\Omega$, then this component will ultimately become negligible compared to the desired correction derived from the PA nonlinearity. This required "smallness" is, however, much more demanding than may at first be assumed. For example, if the delay is $10°$, referred to the modulation period, then the amplitude of the offending sinewave will be approximately 1/5, from (4.6). If the PA is characterized by a third-degree power series $[a_n]$, with the linear gain a_1 normalized to unity, and the input 1-dB compression point defined to be unity signal voltage units, the error signal at the 1-dB compression point is 0.109, much lower than the offending sinewave amplitude. An error signal of amplitude 1/5 would correspond to a voltage gain compression of (1/0.8), or just about 2-dB gain compression. As discussed in detail in Chapter 6, this will be typically beyond the point at which the output can be restored to the correct linear value by increasing the drive level. Such a delay will therefore render the feedback loop to useless, and probably harmful, operation. Even a delay of $1°$, in modulation domain time, corresponds to the amplitude error signal at a compression level of 0.2 dB.

This is a highly significant result in the context of the discussions in this chapter on PA latency. For a modulation envelope having a 1-MHz bandwidth, the $1°$ limit would mean that the PA latency would need to be less than 3 nsec, an unlikely figure in current high-power PA designs. It seems that the use of a compensating delay line on the input detector is *mandatory in all envelope feedback systems*, yet this does not seem to be universal practice. One has to speculate that this may be one reason for the lackluster reputation that this class of linearizer seems to have, and yet this particular problem has this easy "fix," through the use of a compensating delay on the input detected signal. It must be taken as a critical design goal in any such system, that the signals reaching the differential amplifier must be *accurately time-coherent*, down to the subnanosecond range.

Returning to (4.5), the generalized problem really does defy direct symbolic analysis, and the effect of the delay has to be evaluated using a numerical simulation. There are many system simulators now on the market which can tackle such a task. Here we look no further than the venerable SPICE package, but with some tricks employed to keep the analysis entirely in the envelope domain. The block diagram of the SPICE analysis file is shown in Figure 4.12. Since the system works in the envelope domain, it is unnecessary to include the complexity of a modulated RF sinewave, and the excessive computing time this would entail. From the system viewpoint, all of the functions can be specified entirely in the envelope, or modulation domain. Starting with a sinewave generator, the basic two-carrier system can be simulated, with the selected frequency corresponding to the RF carrier spacing. The input and output detectors are assumed to be ideal envelope detectors; this means that the negative cycles of modulation must be rectified. This has been implemented in this simulation using controlled switches.

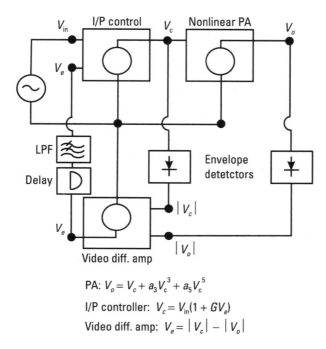

PA: $V_o = V_c + a_3 V_c^3 + a_5 V_c^5$

I/P controller: $V_c = V_{in}(1 + GV_e)$

Video diff. amp: $V_e = |V_c| - |V_o|$

Figure 4.12 Block diagram for SPICE simulation of envelope feedback system.

The RFPA itself can be modeled using a polynomial voltage source, which is another standard element in SPICE. Some care must be taken in choosing the PA coefficients, as illustrated in Figure 4.13. A simple third-degree characteristic can be used for inputs up to the peak value of the cubic function, but for drive levels beyond this point the PA displays a "bistable" characteristic, which is guaranteed to cause computational problems in a feedback circuit of this kind. In open loop use, it is a simple matter to restrict the input drive to levels below the point where the model becomes invalid, but in a feedback loop the drive level is an internal value set by the system. It is therefore more appropriate to consider a model having a higher-degree characteristic, as shown in Figure 4.13, which has a more realistic saturated region. Note also that in the higher region of invalid drive levels, the device still retains a monotonic response. This is a useful safety precaution which reduces the tendency for computational "crashing."

Another important element which is required when using non-zero delay settings is a lowpass filter in the feedback loop. In practice the loop bandwidth will likely be set by the detector and amplifier responses, but in a simulation it is necessary to give the loop an amplitude rolloff characteristic

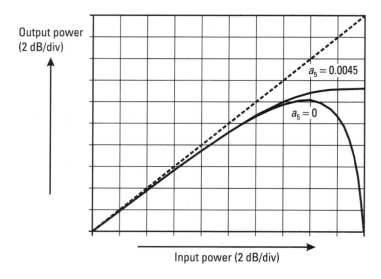

Figure 4.13 PA nonlinear characteristics; dangers of "bistable" third-degree model for simulation.

to keep it stable. The controlling element, or attenuator, also uses a polynomial voltage source, in this case configured to perform the PA input signal scaling and video gain function (Figure 4.14). An adjustable delay is realized using a simple lowpass LC network. This is a convenient and realistic model for a practical video amplifier configuration, incorporating the frequency response associated with the detectors as well as the video amplifier and gain control networks.

Initially, the simulation can be checked out with the delay set to zero. A set of simulated results is shown in Figure 4.15. These plots effectively confirm the quasi-static analysis shown earlier (see Figures 4.8 and 4.9) and show once again the elegant simplicity and effectiveness of feedback techniques for linearizing electronic devices. Implementation of delay *uses equalization between the input and output detectors*; the delay is assumed to be the sum of the delays in the PA, detection, and video processing, and is placed between the detectors and the PA amplitude control element. For a delay corresponding to one-tenth of the modulation period, $\Delta = T/10$, the third-order IM products are hardly shifted from the zero delay case, but the fifth-order IMs show a substantial degradation. Higher values of delay show a more rapid degradation of linearizing performance, and the $\Delta = T/10$ value would seem to be a practical limit for design purposes. This would, for example, correspond to a video delay of 10 nsec in a system having a signal

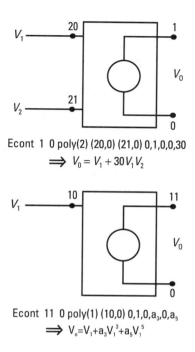

Econt 1 0 poly(2) (20,0) (21,0) 0,1,0,0,30

$$\Longrightarrow V_0 = V_1 + 30 V_1 V_2$$

Econt 11 0 poly(1) (10,0) 0,1,0,a_3,0,a_5

$$\Longrightarrow V_o = V_1 + a_3 V_1^3 + a_5 V_1^5$$

Figure 4.14 SPICE models for PA and input controller.

bandwidth of 10 MHz, although even with this amount of delay the linearization effectiveness can be seen to be restricted for fifth and higher degrees of nonlinearity. This would appear to be a tough, but possibly achievable, goal which would have useful applications in multicarrier and broadband spread spectrum communications systems.

It should be noted that the simulations of Figure 4.15 include a low-pass filter which restricts the linearization action at higher harmonics of the modulation frequency. The IM5 data show very limited improvement, and even substantial degradation, at lower drive levels. This is due to the low starting level of IM5 in the PA itself, and indicates a need for higher video gain if linearization action is still required at these levels. This in turn increases the possibility of oscillation, and would require a lower video bandwidth setting for stable operation at higher values of delay. This vicious circle causes feedback systems to display rapidly degrading performance at higher order IM frequencies. This is a problem frequently observed in practical systems, and relates directly to the loop delay value.

Figure 4.16 shows some of the simulated envelope waveforms for intermediate video gain (30), and $\Delta = T/10$ video delay. Careful scrutiny of these

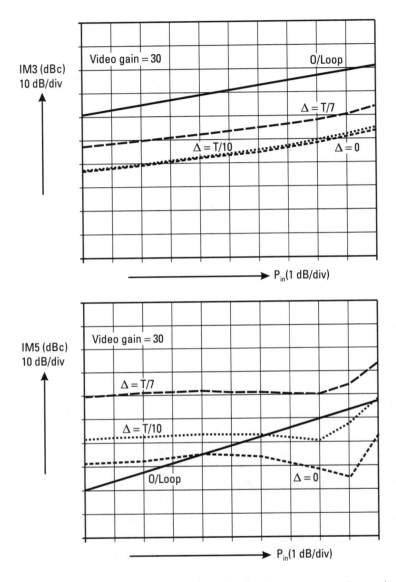

Figure 4.15 Simulation of amplitude envelope feedback loop two-carrier excitation, envelope period T.

waveforms shows that the feedback loop, even in the presence of a delay, still strives to force the output to replicate the input. In a sense, this delay configuration is interpreted by the loop as an additional nonlinearity, or imperfection in the amplifier itself, and seeks to correct it. The presence of the

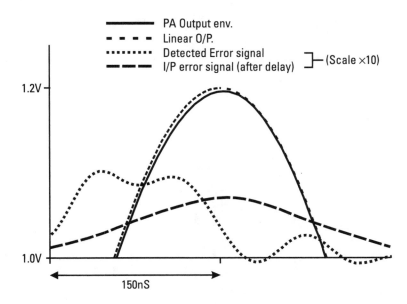

Figure 4.16 Envelope linearization loop waveforms ($\Delta = T/10$ case, $G = 30$, 1-dB compression point).

delay, provided it does not corrupt the error detection process, does not alter the fundamental goal of the feedback system but reduces its effectiveness in executing it. The error signal waveforms also illustrate an important issue concerning the stability of such systems. The loop signal delays have two forms: bandpass delays related to the lowpass frequency cutoff of the various components, and transmission delays caused by cabling and the PA delay itself. Whatever the overall delay, it is essential that the frequency response of the closed loop has a built-in lowpass characteristic which eliminates higher harmonic frequencies where the delay can cause positive feedback. The circuit simulation file in this example uses a lumped element filter which achieves this goal, but will clearly limit the linearization capability at higher envelope harmonics. If this filter is replaced by a simple broadband delay element with similar delay but no frequency cutoff, the simulator fails to converge, indicating a stability problem.[5]

Figure 4.17 shows the corresponding PA input signal, after the application of the "correction" by the input attenuator. The spectral content of the

---

5. In some respects the function of this band limiting has parallels with the video filter in a phase-locked loop.

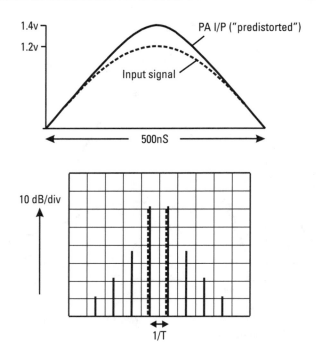

Figure 4.17 Envelope waveform and spectrum of "corrected" input signal to PA (dotted lines show signal input).

PA input envelope shows that the action of the feedback loop is quite similar to that of a predistorter, and many of the results which will be derived in Chapter 5 for predistortion will apply.

In summary, this section has discussed the use and configuration of amplitude envelope feedback in RF power amplifiers. The discussion, analysis, and simulations have shown that such a method of linearization has enormous potential if the delay problem can be resolved. There are many existing applications, particularly single-carrier mobile transmitters, which may have signal bandwidths less than 100 kHz, where the quantitative limits for acceptable delays established in this section must surely be quite feasible. Multicarrier and broadband spread-spectrum signals having bandwidths in excess of 1 MHz clearly pose problems in reducing delays to allow effective and stable higher-degree linearization. But the designer must be aware of the important issues surrounding the detailed design of the gain control element and the use of delay equalization in order to obtain useful results.

There is, however, one major area which remains to be discussed, that of RF phase control.

4.4   Vector Envelope Feedback

Most RFPAs exhibit substantial amounts of AM-PM distortion, and this can in some cases be an equal contributor to the final spectral and demodulation distortion. Clearly, the amplitude-controlled envelope feedback system described in Section 4.3 is insensitive to RF phase, and will not give any beneficial correction of AM-PM effects. Indeed, one of the most common problems is that the amplitude control element may introduce "parasitic" phase modulation. Whether or not an additional phase correction loop is employed, it is of the utmost importance that the amplitude control element has negligible differential phase shift as the attenuation (or gain) is adjusted over its normal operating range. "Negligible" here means essentially unmeasurable; even a few degrees of phase modulation can seriously degrade the improvement from the envelope amplitude feedback control.[6]

In order to include AM-PM correction using envelope domain feedback, some form of RF differential phase detector is required. The most obvious candidate is a simple multiplier, which can be most easily realized using the square-law response of a diode. Basically, a packaged low-cost mixer will usually do the job, provided the IF output frequency range is suitable. There is an immediate problem with this approach, however, in that such an element will also respond to the AM signal variations. This problem is worthy of some simple analysis. Assuming an RF input signal having the form

$$v_i(t) = A(t)\sin(\omega t + \Phi(t))$$

then the output signal can be written in the form

$$v_i(t) = (1 - \delta)A(t)\sin(\omega t + \Phi(t) + \Delta)$$

where δ and Δ represent the gain compression and AM-PM at the envelope input level $v_i(t)$.

So the output of a simple multiplier will be

$$v_m(t) = (1 - \delta)A(t)A(t)\sin(\omega t + \Phi(t) + \Delta)\sin(\omega t + \Phi(t))$$
$$= (\tfrac{1}{2})(1 - \delta)A(t)A(t)\{\cos(\Delta) - \cos(2(\omega t + \Phi(t)) + \Delta)\}$$

which will be reduced by the IF filtering to the video signal,

---

6. This effect is commonly observed as asymmetry in the spectral response.

$$v_m(t) = \left(\tfrac{1}{2}\right)(1 - \delta)A(t)A(t)\cos(\Delta) \qquad (4.7)$$

So if we wish to use this signal to provide drive to a correcting phase adjuster on the PA input the problem discussed initially is now apparent in mathematical form; it is not so much that the differential phase detector output contains an amplitude product of the input and output signal amplitudes, but that the phase detection term is a cosine which has a maximum value at the zero differential phase condition. Such a detector will be unable to distinguish between leading or lagging AM-PM. This problem persists when both input signals to the multiplier have the same phase orientation (i.e., regardless of whether sine or cosine is assumed). What is required is a phase detector giving a sine output function, so that phase lead and lag are distinguished. This can be achieved by using input signals to the multiplier which are time coherent but have a 90° phase.[7] It has already been shown that input and output detectors must work with time-coherent signals, and this will usually be achieved through the use of an equalizing delay line on the input detector; this will apply to both amplitude and phase detection functions. The phase offset could be realized with a simple quarter-wave transmission line phase-shifter in narrowband applications, as shown in Figure 4.18. As with the amplitude correction circuitry, a key requirement is that the system has a well-behaved characteristic in well backed-off, or low envelope points; the phase-shifter must have a characteristic that includes the zero drive condition. It also

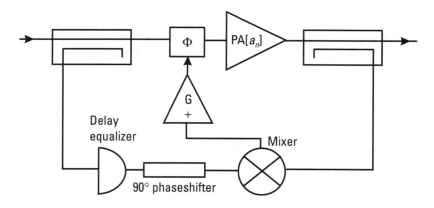

Figure 4.18 AM-PM correction loop.

---

7. This may have already been done by the manufacturer in a commercial phase detector product.

has to be assumed that the amplifier AM-PM will not exceed 90°, when this kind of phase detector reaches a limit in its monotonic behavior.

Returning to the issue of a phase detector which contains some amplitude response in its characteristic [as shown in (4.7)], this will again be unimportant provided that the feedback control loop seeks to minimize the phase difference; the zeroing of the sine function will dominate the characteristic in the range where the differential phase is a small angle. Unfortunately, the assumption that a mixer of conventional design can behave as an ideal multiplier over a wide dynamic range of equal amplitude input signals on both input ports is somewhat optimistic. Most mixer designs based on diode elements require the local oscillator (LO) port to receive a signal of fixed and substantial amplitude, causing the diodes to behave as switches. Although mixers optimized for use as phase detectors are commercially available, there is still a dynamic range issue in using this approach.

An alternative approach, which may alleviate some of the problems associated with RF envelope detection, would be to use conventional mixers to convert the input and output signals down to baseband, where gain and phase detection could be performed using more robust analog or DSP techniques. Such a system may appear to be re-inventing the Cartesian Loop, but in fact occupies a potentially useful halfway position between an RF vector envelope feedback system and a baseband feedback system. In particular, it requires only the RF carrier information, not the entire modulation function. The downconverters can be supplied with an LO signal of optimum strength, and most critically, will provide baseband outputs which have the same bandwidth as the original RF signal. Such a system would, inevitably, introduce more delay into the feedback loop.

An additional cause of delay in a PA feedback system is in the video processing components, especially the high-gain video amplifiers which follow the phase and amplitude detectors. Indeed, given that it may be impractical to reduce high-power PA delays much below 10 nsec using current device technology, the video delays may appear to be the dominant source of performance degradation. It has already been observed that just about any attempt to design and build an envelope linearization loop will require special detectors to be designed; typical connectorized lab detectors will usually be too slow for these applications.

Video gain, extending from dc up to perhaps 10 MHz or more, appears to pose problems in terms of latency specs available in commercial video amplifiers. This, however, may be a misconception based on traditional views about gain at low frequencies. The key issue in the present application is that the linearity requirements of the video amplifier can be greatly relaxed.

A gain block at these frequencies has become so closely associated with an operational amplifier configuration that the possibilities offered by simple discrete transistor amplifiers are widely forgotten. The key point is that because the video amplifier is part of a feedback system, its own linearity is as uncritical as that displayed by the much higher open loop gain of a typical operational amplifier. But the actual gain requirement here is so modest that very few stages may be required and it may be more appropriate to use discrete, fast RF devices in a simple video gain configuration. This is an attractive possibility for RFIC implementation.

This section has laid out some proposals, and indicated possible hardware solutions, to the realization of full-vector envelope feedback around an RFPA. The elimination of downconversion to baseband or IF, as required by more traditional schemes, allows more compact circuitry and lower video delays which greatly extend the linearized signal bandwidth range. It also has interesting and practical attractions for implementation in RFIC and hybrid module technologies.

4.5 Low Latency PA Design

A typical high-power 2-GHz RFPA, measured from input to output socket, will have a delay in the range of 5 to 30 nsec. A substantial fraction of this delay will come from high-Q matching networks, mainly those in the high-power output devices. Another important contribution comes from the interconnecting transmission lines, couplers, combiners, isolators, and RF control circuit elements. Very frequently, the PA itself will have been designed by a separate group, or even purchased as an off-the-shelf item, and the delay will not have been regarded as a critical specification item. Indeed, even if such specs exist, they may well be based on a "what-it-does-is-what-you-get" basis. The fact is this single parameter makes a direct statement on the signal bandwidths for which feedback linearization is feasible. Even a reduction of a factor of, say, 2 in the PA delay will effectively double this bandwidth. A factor of 5 or 10 could take the linearizable signal bandwidth into a multicarrier regime. The issue has to be attacked on both fronts; reducing the Q-factor of the matching networks, on one hand, and reducing the interconnection delays, on the other.

The Q-factor issue for high-power devices is essentially technological. RF transistors are low-voltage and high-current in nature, and this means low loadline impedances on the output. Input impedances are also typically highly reactive, leading to the need for high-Q resonators in order to extract optimum gain. The third part of this equation is the ubiquitous 50-Ohm microwave environment. Clearly, the output port of an amplifier has to be 50 Ohms, to interface with other parts of the system, but surely the interstage matching can be done at any level the designer chooses? This is certainly worthy of further quantification, although it should be stressed that it is just as much the reactive part of the RF transistor impedance that causes bandwidth and latency problems than its low real part. There is no question, however, that if an RF transistor technology came along that had higher voltage, and consequently higher loadline impedance, along with lower or comparable normalized parasitics, there should be a strong interest from the linearized PA design community. Such new technology seems to be well on its way, in the form of Silicon Carbide (SiC) and Gallium Nitride (GaN) FETs.

With such developments already in progress, it is instructive to consider PA delays in a quantitative manner. A good starting point is to look at a basic microwave-matching network from the viewpoint of phase delay. Such a circuit is shown in Figure 4.19. In *RFPA* this circuit was described as the "Occam's Razor" of microwave matching; the simple lowpass network is widely used, either in single or multiple form in order to match the low impedances of RF power transistors up the 50-Ohm level.[8] A typical scenario would be a 50:1 impedance ratio; RF power transistors can quite typically have impedances, both input and output, in the 1-Ohm range.

Using the design equations (see *RFPA*, p. 76) for the single lowpass network in Figure 4.19, the impedance transformation ratio, m, is given by

$$m = \frac{R_O}{R_T} = 1 + \left(\frac{R_O}{X_C}\right)^2 = 1 + Q^2 \qquad (4.8)$$

where Q is the overall *Q-factor* of the network.

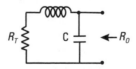

Figure 4.19 Lowpass matching network.

---

8. One might ask, and many do, why 50 Ohms? The answer is, long-established convention; although some interstage matching between power devices may in principle be performed at lower impedance levels, the higher-power RF devices tend to be pre-matched to 50 Ohms inside their packages.

So for a value of $m = 50$, corresponding to a 1-Ω to 50-Ω transformation, $Q = 7$. This results in a 3-dB bandwidth of approximately $\omega_o/7$, or just over 10% centered around the resonant frequency ω_o. Of greater interest here is the complementary result that the slope of the phase characteristic of such a network around the resonant frequency is given by the simple relationship

$$\frac{d\phi}{d\omega} = \frac{2Q}{\omega_O} = \frac{QT}{\pi} \qquad (4.9)$$

where T is the RF cycle period.

For the purposes of the present discussion, we will define this term as the "latency" of the network. It has the units of time, and gives a direct reading on the delay of the network, measured in terms of a single RF cycle period T. For example, the 1-Ω matching network at 2 GHz will have a latency of Q/π, or just over two RF cycles (1.1 nsec).

So this simple matching network, which may occupy a small space inside the transistor package, has an effective electrical length of just about 30 cm (1 foot). This is just one network; in practice both input and output will have similarly low impedances, so that a high-power transistor, matched to 50 Ohms, will quite typically have an electrical length of over four RF cycles. At higher power levels, this length could increase substantially, as yet more cells are paralleled to obtain more power. Clearly, such a device, after matching, poses a major problem for feedback design. Not only will the rapid phase rotation, quantified by (4.9), cause the device to be in positive feedback mode at a fractional frequency of $7\omega_o/8$, but also at even closer frequencies, the benefits of the negative feedback will be lost.

There is, unfortunately, another effect which can multiply the individual matching network delays in a typical higher power RF gain stage, which is due to the non-zero value of the reverse transmission coefficient, s_{12}. Traditional teaching asserts that providing the stability factor k is greater than unity, a device can be conjugately matched and unconditional stability can be obtained. This simple result conceals a complicated matching scenario, in which both input and output networks interact with each other and can significantly boost the overall gain and Q-factor as compared to a unilateral device having the same matching and forward transmission parameters. This applies in situations where the k-factor can be substantially in excess of unity.

Figure 4.20 illustrates this in a specific case, using a commercially available 2-GHz, 10-W RF power transistor. Based on the input and output mismatch (s_{11}, s_{22}) and the forward gain (s_{21}), the unilateral version of the device (i.e., obtained by setting $s_{12} = 0$) would be given by

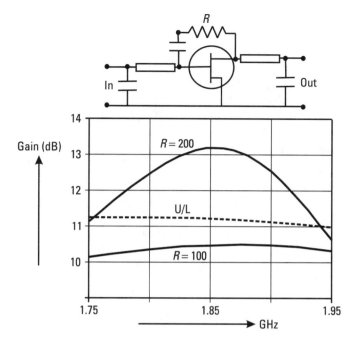

Figure 4.20 PA designs of 10W using different *k*-factor selections: *R* = 200 (*k* = 1.04), *R* = 100 (*k* = 1.2), and "U/L" unilateral case, $s_{12} = 0$.

$$G_u = \frac{|s_{21}|}{\left(1 - |s_{11}|^2\right)\left(1 - |s_{22}|^2\right)}$$

which would be about 12 dB in this case. The overall Q-factor of the gain response, shown in Figure 4.20, is a direct function of the input and output matching networks, as is the phase-frequency characteristic, shown in Figure 4.21.

If we now restore the reverse transmission *s*-parameter coefficients to their data sheet values, the matching networks have to be modified. Noting that the *k*-factor is lower than unity, a resistive element has to be used in order to make the device unconditionally stable over the band of interest (see inset in Figure 4.20 for schematic). We can now use the classical conjugate matching formulae found in any elementary RF textbook, to retune the matching networks. Logically, we will set the damping resistor to the highest possible value, which results in *k* > 1, since lower values result in substantial gain reduction. The results, also shown in Figures 4.20 and 4.21, put a rather

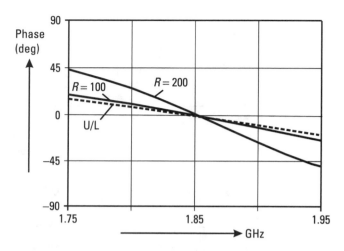

Figure 4.21 Phase response for 10-W PA designs.

different perspective on the conjugate matching issue than is usually given in the textbooks. Although the device is unconditionally stable, the final gain and Q-factor show considerable increase as compared to the ideal unilateral device, depending on the choice of the "damping" resistor. A lower value of damping resistor further reduces the overall Q-factor and has a substantial effect on the slope of the phase characteristic. This third design, having a device whose *k*-factor has been overcompensated from the stability viewpoint, clearly offers an important option for the designer: lower gain with lower latency. In fact, in this case, the final in-band gain shows a reduction from 13.5 dB to 11 dB for the two practical cases, while the phase slope is reduced by a factor of about 2.5. In terms of actual delays, the two designs represent values of 111 ps and 44 ps, respectively.[9]

    Clearly, such effects need to be recognized and utilized by PA designers in order to lower the delay in RFPA gain stages. The loss of an extra decibel or two of gain is possibly an acceptable price to pay for a reduction in delay of maybe a factor of two. This kind of design tradeoff is almost unknown in conventional PA design circles. Obviously, commercial pressures on device

---

9. The reader may be surprised at how small these delays are, considering that data from a commercially available 10-W device is being used; it so happens that the device in question is a Silicon Carbide device. It should also be noted that in practice the output will not be conjugately matched; this does not, however, affect the overall Q-factor issue significantly.

manufacturers tend to push them in the "bigger is better" direction as far as power gain is concerned. As device gains increase due to improved, or new, device technology, it may pay to maintain gain at similar levels and focus on latency reduction.[10] Gain can be replaced at lower power levels in the chain, where delays are much lower.

Another prime source of delay in multistage higher-power PA assemblies is the excessive use of physical space. Printed circuit boards, for all their cheap convenience, use up real estate and associated electrical length much more extravagantly than the much tighter packing densities available in integrated circuits and hybrid modules. The use of packaged discrete components is part of the problem, and hybrid assembly technologies based on the use of chip components have for many years demonstrated the possibilities for greatly reducing the size of electronic subsystems.

The application of hybrid technology to the multistage RFPA seems to have received little attention. Higher current densities in thin circuit traces, and thermal dissipation are obvious instant detractions. But if a typical packaged RF power transistor is used as a starting point, some viable possibilities emerge. We already have internal matching elements incorporated inside the package of a high-power device. Why not extend the concept and include a driver stage, greatly reducing the electrical length of the interstage circuitry, as compared to a typical discrete PCB assembly? Then place a vector feedback loop around a gain block, which now has minimum electrical delay. Some of the necessary transistor elements could be placed on a single die, for ease of assembly. As a minimum, such a hybrid component could be offered as a "super-component" building block for higher level, more conventional, integration. If the Q-factor of high-power RF transistors is a real limitation, it seems a viable option to place linearization loops around smaller devices with lower delays, and then use power-combining techniques to reach the required final power level.

The frequently perceived incompatibility of high chip heat dissipation and hybrid technology is largely based on the use of ceramic, especially alumina, substrates. But the use of metal ribs or septums for the mounting of higher power devices in ceramic substrates has been practiced in the microwave hybrid (MIC) (see Chapter 7) industry for many years. Even the synthetic diamond has progressed to the point where it could be considered as a hybrid substrate material. In the final count, if a power device can be

---

10. This level could be conveniently set at about 10-dB gain for an RF power output stage, below which overall efficiency, or power-added efficiency (PAE), will start to degrade more seriously.

supplied in a package, there is no fundamental reason that the same device cannot be supplied in a somewhat larger package, using the same technology, but containing a higher level of integrated components in the form of driver stages, gain stages, and linearization circuitry.

4.6   Variations

There are numerous variations which can be proposed for envelope feedback systems, once the basic principles of viability have been established. This section lists a few of these, without any detailed attempts to analyze or simulate any particular configuration in detail.

One important class of systems which has not been considered is the use of power control in the PA output, rather than using an input gain control. Such a system, shown in Figure 4.22, has some distinct advantages over conventional input control. In particular, it is free from the limitations of predistortion; in principle, there is no limit to the amount of control that can be applied, and the control does not itself introduce additional higher-degree effects (see Chapter 5). There is, of course, a price to be paid in the form of the available output power. If the controller is still a simple voltage-controlled attenuator, there will be a residual attenuation setting of maybe 2 dB or so in the linear region, in order to provide suitable linearization range.

This loss is certainly undesirable in a high power amplifier, although it should be fairly pointed out that similar losses are inevitably incurred in the output of a PA in a feedforward system, when the insertion coupling factor and delay line losses are added up. It can even be argued ([6]; see also

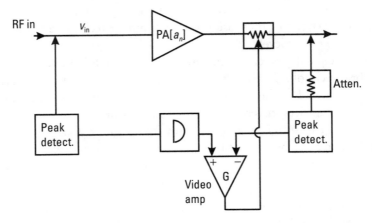

Figure 4.22 Envelope feedback using output power control.

Section 6.4) that the negative impact of attenuation in the output of a linearized PA has to be considered in relation to the positive benefit of the linearization system; if a PA running at 1-dB compression has its output power boosted back up to the uncompressed level by the linearizer, the system as a whole breaks even on power if the attenuation can be kept below 1 dB. In the case of an output attenuator power control, the power wastage will always be on the negative side, depending on the minimum attenuation which can be achieved. It should be noted further that any nonlinearity in the attenuator itself will in principle be removed by the feedback system. An additional important benefit of this system is the reduction of loop delays down to just the video processing components. If the signal from the input detector is delayed, in the manner already described, so that it is time coherent with the detected output signal, the correction signal emerging from the video differential amplifier is now being applied to the PA output, thus eliminating the requirement for very low latency in the PA. The time discrepancy in this system is essentially down to the delays in the detection process, the video gain, and the attenuator drive circuitry.

Realistically, such a system is unpopular due to the use of a large lossy element in the PA output. Rightly or wrongly, this seems to have firmly stamped this approach as a dead end. It is therefore worth considering alternatives for the output control device, given that the other attributes of such a system are positive. One such possibility takes yet another page from the feedforward book, whereby an additional amplifier is used to provide the output correction signal, Figure 4.23. This is additive correction, and need

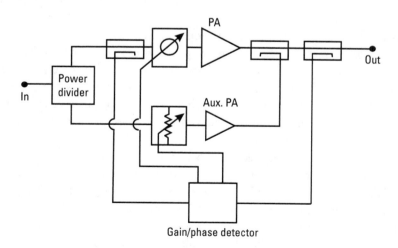

Figure 4.23 Envelope feedback system using auxiliary PA to linearize output.

not be the sole preserve of feedforward systems. The amplitude error signal, obtained as usual using detectors and suitable time delay correction, is now used to control a second auxiliary amplifier. The action of the control loop will cause the input attenuator on the auxiliary amplifier to adjust to a suitable level which neutralizes gain compression in the main PA. It will be necessary to have a backed-off, or linear, condition set up such that some power is being supplied by the auxiliary amplifier, so that any gain expansion can be corrected by further reduction in the auxiliary amplifier contribution.

One might well ask, looking at Figure 4.23, why not simply build a feedforward system?

One answer to this question could be that feedforward systems are not "simple." The system of Figure 4.23 is closed loop and essentially self-adapting. This incurs the hazards of loop delay effects, but eliminates the need for housekeeping overheads such as precision gain and phase tracking adaption. There is also a more favorable situation regarding the power requirements of the auxiliary amplifier. The closed loop adjusts the power from the auxiliary amplifier on a dynamic basis, and the linearity requirements for the auxiliary amplifier are minimal. The auxiliary amplifier power level would have to be sufficient to overcome the output insertion coupling factor, as in a feedforward loop (see Chapter 6), but the amplifier itself could be considerably more efficient than its feedforward counterpart.

The necessary inclusion of a phase control loop, also shown in Figure 4.24, provides some relief on the power level required from the auxiliary amplifier in comparison to a feedforward system. As will be discussed in some quantitative detail in Chapter 6, the use of output power addition to correct AM-PM distortion in a feedforward system is a wasteful process, and substantially increases the error PA requirements. In the present system, the AM-PM is corrected in a separate loop, using a simple phaseshifter. This system will run into loop delay limitations which will ultimately reduce its ability to correct higher-degree nonlinearities. But it possibly represents the optimum configuration for using feedback methods to achieve comparable levels of linearization as can be achieved using more popular, but cumbersome, open-loop techniques.

It was mentioned in Section 4.5 that the challenging requirements of PA latency for successful envelope feedback implementation can be substantially reduced if smaller PA modules are combined. Smaller devices have lower Q-factors; indeed, for a given device type, the Q-factor will scale down in a linear fashion as the power, or number of cells paralleled on the die, decreases. So if a 1-nsec PA delay looks challenging in a multicarrier 3G application requiring a 10-MHz signal bandwidth, an immediate relaxation of a factor of 8 can in principle be obtained by combining eight linearized modules, as indicated in Figure 4.24. Note that in order to obtain the full factor of 8, it is necessary to linearize each individual power module; some benefit could still be obtained by using a single linearization loop around the complete system, but the factor-of-8 benefit would be reduced by the electrical length associated with the power combiners. An additional reduction in complexity could be considered, and is shown in Figure 4.24, whereby the control signals for each individual gain/phase linearizer could be derived from a single module having the feedback loop. If the other modules are sufficiently similar in their characteristics, they could be linearized using a common drive signal applied open-loop fashion.

Finally, and looking ahead to Chapter 5, an envelope feedback system can be used as the basis for a digital predistorter, as shown in Figure 4.25. This is a possible way to outflank, if not defeat, the problem of loop delay. The system is run initially in a conventional envelope feedback mode, but at a low modulation frequency where the loop delays have a negligible impact on the linearization performance; this may, for example, be in the 10–100-kHz region. Typically, a representative signal environment would be used, not necessarily a simple two-carrier system. The control voltages generated at the gain and phase modulators can be read by an analog-to-digital converter (ADC), along with the corresponding envelope amplitude levels, thus

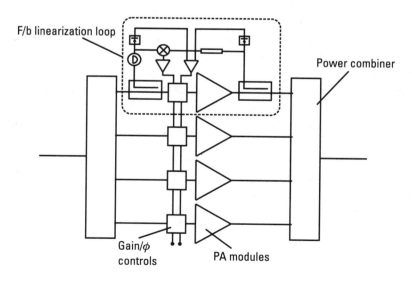

Figure 4.24 Power-combined modules using feedback linearization control derived from a single element.

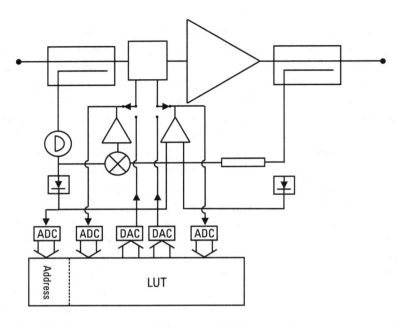

Figure 4.25 Envelope feedback used as a basis for LUT calibration.

establishing a look-up table (LUT) for the required settings of gain and phase correction under a given signal environment. The system can then be switched into open-loop operation, where the DAC now recreates the drive signals to the modulators, based on the reading of the input envelope amplitude. So long as the PA linearities are quasi-static in nature, the corrections can now be applied at a rate which can be as fast as the DSP components can allow. The system is now a predistorter, but it uses information obtained in a closed loop feedback configuration to generate the predistorted signal. As discussed in Chapter 3, memory effects will come into play and it cannot be assumed that the corrections will remain valid, and the integrity of linearization maintained, at modulation frequencies which are orders of magnitude higher than the LUT calibration data. But the LUT data will certainly be much closer to the required dynamic values than if based on a simple static measurement of the PA gain compression and AM-PM response.

4.7   Conclusions

This chapter has laid out a strategy for the design of RF power amplifiers using feedback linearization that recognizes, but tackles, each and every objection that is typically and conventionally leveled against such an approach. It is worth summarizing these issues:

1. Use of custom-designed high-speed amplitude and phase detectors to sense gain compression and AM-PM; elimination of downconversion to perform these tasks;

2. Use of input delay line to achieve time coherence on input and output detection processes;

3. Use of gain and phase control elements which have well-behaved responses when PA is in linear, or well backed-off, condition;

4. Careful tailoring of the frequency rolloff of the feedback loop response to be compatible with the overall feedback delay;

5. Design of low-latency PAs; use of newer, high-voltage PA device technology (SiC, GaN), hybrid circuit technology, power combining of lower power feedback modules.

This is quite a long list, and almost every item requires its own development program. Such effort must be compared with the large development times associated with complex adaptation methods which are mandatory in open-loop systems.

References

[1]   Alexander, R. C., *Inventor of Stereo: The Life and Works of Alan Dower Blumlein*, Woburn, MA; Oxford, England: Focal Press, 1999.

[2]   Blumlein, A., and H. Clark, U.K. Patent 425,553 ("Amplifier using negative feedback loop"), September 1933.

[3]   Black, H. S., U.S. Patent 1,686,792 ("Translating system"), October 1928.

[4]   Kennington, P., *High-Linearity RF Amplifier Design*, Norwood, MA: Artech House, 2000.

[5]   Seidel, H., "A Microwave Feedforward Experiment," *Bell Syst. Tech. Jour.*, Vol. 50, November 1971, pp. 2879–2916.

[6]   Seidel, H., H. R. Beurrier, and A. N. Friedman, "Error-Controlled High Power Linear Amplifiers at VHF," *Bell Syst. Tech. Jour.*, Vol. 47, May–June 1968.

5

Predistortion Techniques

5.1 Introduction

Predistortion is a useful method of achieving linearity improvement in RF power amplifiers. Its appeal can, however, all too easily exceed its ultimate performance. The concept of placing a small, magic box on the PA input, which consumes little power and provides linearization comparable to more complex methods such as feedforward, is compelling but also naive. Fundamentally, all predistortion methods are open-loop and as such can only approach the levels of linearization of closed-loop systems for limited periods of time, and over limited dynamic range. Predistortion (PD) methods have nevertheless, been the focus of much recent research and development, mainly due to the renewed possibilities offered by DSP. But the value of predistortion largely remains as a complementary technique working in tandem with a system using feedback or feedforward. In particular, as will be analyzed in Chapter 6, the use of a well-designed predistorter on the main PA in a feedforward loop can substantially reduce the power requirements of the companion error amplifier, thus resulting in a significant increase in overall efficiency. There are also some applications, for example mobile transmitters, where the simplicity and almost zero cost of a simple predistorter is well worth the few decibels of ACP or IM reduction that can be obtained over a limited power range. The realization of PD-only PA systems which can truly compete in performance with more conventional feedforward techniques in MCPA applications is an active, but still largely unfulfilled, research area.

One of the main goals of this chapter is to put predistorter design on to a firmer, *a priori* basis than has typically been the case in the past. Simple analog predistorters have been all too often the result of empirical adjustment of a simple circuit, usually incorporating a diode or two, which has been "jimmied up" to give a rough approximation to an expanding gain characteristic. Such efforts, which are still to be seen in the literature and on the symposium circuit, usually result in a combined PD-PA characteristic which shows a deep notch in the two-carrier IM3 response at a drive level near to the 1-dB compression point of the PA. Closer scrutiny of the measured data on such efforts will usually show a number of less desirable characteristics: much less improvement in higher order IMs (even showing degradation in some cases), and a substantial "in-filling" of the IM3 notch in multicarrier or spread spectrum signal tests.

The design methodology developed in this chapter is based firmly on the PA nonlinear modeling methods discussed in Chapter 3. The first step is to perform the mathematical inversion of the PA Volterra series in order to establish the required characteristics of the PD itself. This process, in itself, results in some very useful general principles about the limits of PD performance, and can be used to explain the "notching" behavior that is frequently observed. The second step is to consider various methods for synthesizing actual PD configurations which have the prescribed characteristics. Both analog and DSP implementation are described, but both use the same fundamental recipe derived in the first part of the chapter.

Conceptually, a predistorter is appealingly simple, and is illustrated in Figure 5.1. A typical PA gain compression characteristic is shown, which for simplicity is assumed to have a simple third-degree nonlinearity. The action of the predistorter, at any typical input signal level, is shown by following the signal paths on the PA and the corresponding extrapolated linear characteristic. For an input signal at level V_{in}, the amplifier shows some compression, resulting in an output level V_a; an ideally linear amplifier would give an output shown as V_o. In order to give this linear output, the action of the predistorter is to increase the input level V_{in} to a higher level V_p, which can be obtained graphically from the gain compression characteristic by drawing a horizontal line from the intersection of the V_{in} level with the linear characteristic (point "A" in Figure 5.1) across to the intersection with the actual amplifier characteristic (point "B"), then down to the horizontal axis to determine the required PD output level, V_p.

Before casting this simple graphical concept into more concrete mathematical terms, it is worth making a few observations which should be remembered throughout all that follows in this chapter.

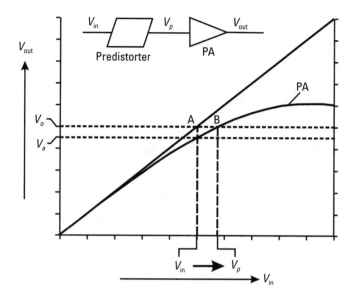

Figure 5.1 Basic action of predistorter.

1. Predistortion, in a sense, attempts an intuitive contradiction: As a device distorts, alleviate the distortion by driving it harder.[1]

2. The process clearly runs into difficulties as the amplifier saturates. There is a point of no return, where no further increase of drive level can restore the output to the desired point on the linear characteristic. This issue assumes a much greater significance in modern communications systems, where signals having high PEP to average power ratios are commonly used.

3. The signal emerging from a predistorter *will be highly distorted.* Indeed, the signal emerging from an effective PD, viewed on a conventional spectrum analyzer, will display very similar spectral distortion to that which would be observed on the signal emerging from the uncompensated amplifier. This observation can have a critical impact in terms of the required bandwidth of the PD and/or the speed of the DSP circuitry; it may also have more far-reaching implications as high-speed data signals spread to fill up the entire amplifier bandwidth.

---

1. This surely triggers some recollections of the humorous sign posted in many an office area, "the beatings will continue until morale improves."

4. The predistorter is shown here as having "gain." In practice, the PD will usually be a passive device, whereby the "gain" is achieved by a reduction in the PD attenuation. This does not fundamentally affect the analysis or conclusions, and the convenient assumption of a PD with gain will be used in the ensuing analysis.

5.2 Third-Degree PA: Predistortion Analysis

We will return to all of the above issues in due course, but now perform some analysis on the simple third-degree PA predistorter scheme shown in Figure 5.1. Lowercase voltage symbols are now used to indicate functions of time,

$$v_o = a_1 v_p - a_3 v_p{}^3$$
$$= a_1 v_{in}$$

so that

$$v_p{}^3 - \left(\frac{a_1}{a_3}\right) v_p + \left(\frac{a_1}{a_3}\right) v_{in} = 0 \tag{5.1}$$

So the required functional relationship between the input signal level v_{in} and the predistorter output v_p is the root of a cubic equation, in the case of an amplifier characteristic having a simple third-degree nonlinearity.

It would be of some relevance here to reflect on the colorful history surrounding the analytical solution of a cubic equation of this kind, since although a simple iterative algorithm will crunch its way to a solution of adequate precision, some important further insight can be obtained from a solution in analytical form. The required ideal PD characteristic, obtained using such an iterative method, is plotted out in Figure 5.2. Close inspection of the PD characteristic in Figure 5.2 shows that at drive levels well backed off from the compression region, the required characteristic approximates to a complementary gain expander; the PD simply has to provide gain expansion which cancels the corresponding gain compression of the PA. As the amplifier is driven into more substantial compression, however, some extra expansion is required from the PD due to the fact that a given increase in drive to the PA no longer results in a corresponding increase in output. So the required increase in PD output starts to escalate towards a point of no return when the PA reaches saturation.

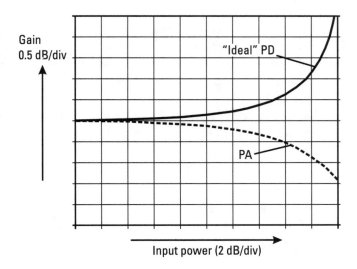

Figure 5.2 Ideal predistorter gain expansion characteristic for PA having third-degree nonlinear gain compression.

Some further useful insight into the possible realization of a suitable PD device can be obtained by expressing the solution of (5.1) in an analytical form. The classical solution of Cardano will prove unhelpful in this case, (5.1) being an example of the so-called irreducible cubic having three real solutions. What we require is a power series expression for the solution in order to establish the PD characteristic; this can be obtained by engaging an *ad hoc* iteration routine in symbolic form. Rewriting (5.1) and setting $a_1 = 1$, the successive application of

$$v_p = v_{in} + a_3 v_p^{\;3}$$

will give a better approximation to the value of v_p. Seeding the iteration with the initial approximation $v_p = v_{in}$ gives an infinite series for v_p in terms of v_{in},

$$v_p = v_{in} + a_3 \left(v_{in} + \; \ldots \; a_3 \left(v_{in} + a_3 \left(v_{in} + a_3 v_{in}^{\;3} \right)^3 \right)^3 \ldots \right)^3 \quad (5.2)$$

$$= v_{in} + a_3 v_{in}^{\;3} + 3 a_3^{\;2} v_{in}^{\;5} + 12 a_3^{\;3} v_{in}^{\;7} + 37 a_3^{\;4} v_{in}^{\;9} + \; \ldots$$

Although we are still using the simplest possible third-degree model for the PA, this result has some quite far-reaching conclusions. Most notably, an ideal PD has to have a power series containing an infinite number of higher-degree terms, despite the initial assumption that the PA has only third-degree nonlinearity. The third-degree PD term has, as might be expected, the same magnitude but opposite sign to the PA power series third-degree term (a_3). But a PD having only a third-degree characteristic will predistort the input to the PA such that higher-degree terms will now be generated by the PA characteristic. For example, if the PD characteristic is expressed as

$$v_p = v_{in} + b_3 v_{in}^{\ 3}$$

then the amplifier output will be

$$
\begin{aligned}
v_{out} &= a_1 v_p - a_3 v_p^{\ 3} \\
&= a_1\left(v_{in} + b_3 v_{in}^{\ 3}\right) - a_3\left(v_{in} + b_3 v_{in}^{\ 3}\right)^3 \\
&= a_1 v_{in} + \left(a_1 b_3 - a_3\right)v_{in}^{\ 3} - 3a_3 b_3 v_{in}^{\ 5} - 3a_3 b_3 v_{in}^{\ 7} - a_3 b_3^{\ 3} v_{in}^{\ 9}
\end{aligned}
\tag{5.3}
$$

So clearly, if we make a PD which has $b_3 = a_3 / a_1$, all third-degree distortion will be removed from the PA output, but there will now be some new additional higher order distortion that was absent in the original PA. In order to remove these higher order products, the PD has to have the additional higher-degree distortion terms shown in (5.2). We will see in due course that quite useful performance can be obtained from a PD that has a heavily truncated approximation to the ideal response.

It is worth commenting at this point that nature probably doesn't work in polynomial form, and that just because a PD device having the ideal characteristic shown in Figure 5.2 has an infinite polynomial series representation does not necessarily mean it cannot be built; curves are curves, however one chooses to model them. There is, however, a very good reason for basing PD theory and design on the polynomial representation, as discussed in Chapter 3: The power series terms all have direct significance in the frequency domain. In particular, the lower degree nonlinear terms, such as third and fifth, are the most troublesome in communications applications and merit closest study. Even in the practical situation of a PA whose characteristics cannot be accurately modeled using just a couple of lower degree nonlinear polynomial terms, the removal of such lower degree nonlinearities will always give a major improvement in PA performance. A power series representation is also most convenient for the synthesis of a desired PD function, either using analog elements or DSP algorithms.

The polynomial expression in (5.3), which represents a third-degree PA and a third-degree predistorter, also indicates a further general issue concerning the properties and classification of predistorters. In Chapter 3 it was reiterated that third-order distortion products, such as third-order IM or adjacent channel spectral regrowth, can arise from degrees of nonlinearity higher than the order of the distortion effect. For example, in (5.3), an IM3 product will be generated by all of the terms except the linear one. Thus, the elimination of third-degree distortion by suitable setting of the PD third-degree coefficient b_3 will not eliminate third-order IM distortion. But the correct value of b_3 will cause the power backoff (PBO) slope of any residual IM3 to be at least 5:1, since the lowest degree of distortion is now the fifth power. Such a PD is termed a "matched" PD, for third-degree effects. If the value of b_3 is not "matched" and has a non-optimum value, there is still a possibility for the now residual third-degree IM3 product to cancel with the IM3 contributions coming from the higher-degree terms. Such cancellation will only occur at a single specific drive level, and the cancellation level will be specific to only a single distortion product. Nevertheless, such cancellation can be useful in some applications, and will be considered further. Such a predistorter will be termed a "notcher." We will see that the matched PD is a much more robust and generally useful device than the notcher, and that the concept can be extended to handle more complex, and practical, PA characteristics than the simple third-degree memory-less system considered here.

It is instructive at this point to evaluate gain compression and two-carrier IMD responses for the third-degree PA with some specific cases of PD response. Assuming a PA characteristic

$$v_{out} = a_1 v_p - a_3 v_p^{\ 3}$$

and normalizing the terms such that $a_1 = 1$, and the 1-dB compression point occurs at an input level $v_p = 1$ (giving $a_3 = 1 - 10^{-0.05} = 0.109$).

Case 1: Third-degree nonlinear PA, PD with matched third-degree characteristic

For a matched third-degree (only) PD characteristic, $b_3 = a_3$, from (5.2), and $b_5, b_7, \ldots = 0$. From (5.3), the composite PD/PA power series now shows a null third-degree term:

$$v_{out} = a_1 v_{in} - 3a_3 b_3 v_{in}^{\ 5} - 3a_3 b_3 v_{in}^{\ 7} - a_3 b_3^{\ 3} v_{in}^{\ 9}$$

and the key point here is that the third-degree term is zero for *all values of the drive level v_{in}*. So for a two-carrier signal, the IM3 distortion will show essentially a 5:1 slope, the higher-degree terms being negligible up to the 1-dB compression value of $v_{in} = 1$.

The PA, PD, and composite gain compression characteristics are shown in Figure 5.3, along with the PA and composite IM3 response.

Case 2: Third-degree nonlinear PA, PD with matched third- and fifth-degree characteristic

For a matched third- and fifth-degree PD characteristic, $b_3 = a_3$, and $b_5 = 3a_3^2$ [from (5.2)]. The composite PD/PA power series now shows a null third- and fifth-degree term, for all input drive levels, but now has additional terms extending up to the fifteenth degree:

$$v_{out} = a_1 v_{in} - 3a_3 \left(b_3^2 + 3b_5\right)v_{in}^7 - a_3 b_3^3 v_{in}^9 - 3a_3 b_5 \left(b_5 + b_3^2\right)v_{in}^{11}$$
$$- 3a_3 b_3 b_5^2 v_{in}^{13} - a_3 b_5^3 v_{in}^{15}$$

so the composite PA/PD IM3 response will now have a 7:1 slope, showing only residual seventh-degree distortion (and higher degrees at a negligibly low level); this is shown in Figure 5.4. In practice, of course, the PA will have some fifth-degree distortion of its own, but this can still be cancelled by an

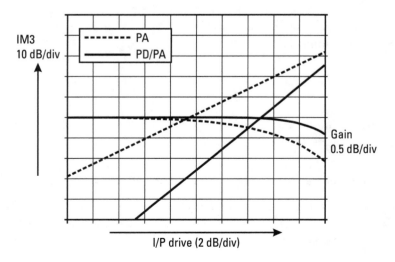

Figure 5.3 PA with third-degree nonlinearity (dotted); composite PD/PA response with third-degree PD (solid; $b_3 = a_3$).

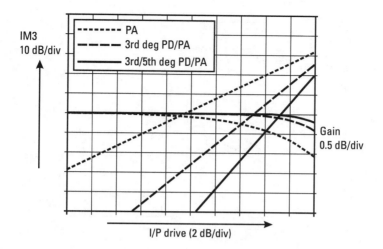

Figure 5.4 PA with third-degree nonlinearity (dotted); composite PD/PA response with PD having third- and fifth-degree characteristics (solid, $b_3 = a_3$, $b_5 = 3a_3^2$).

appropriate readjustment of the PD coefficients to give the same results as in Figure 5.4.

Case 3: Third-degree nonlinear PA, PD with "unmatched" third-degree gain expansion characteristic

Figure 5.5 illustrates a situation where a PD has a roughly matching gain expansion characteristic which in fact only passes through the exact required PD characteristic at a single drive level. This turns out to be a very common practical situation. Although now none of the power coefficients of (5.3) are zero,

$$v_{out} = a_1 v_{in} + \left(a_1 b_3 - a_3\right)v_{in}^{\ 3} - 3a_3 b_3 v_{in}^{\ 5} - 3a_3 b_3^{\ 2} v_{in}^{\ 7} - a_3 b_3^{\ 3} v_{in}^{\ 9}$$

if the third- and fifth-degree terms have opposite sign (e.g., $b_3 > a_3/a_1$), then there will be a single specific level of v_{in} at which the third- and fifth-degree contributions to the IM3 products will cancel. In the simplest case of a two-carrier signal, there will appear to be complete cancellation of IM3 products at this level; the IM3 output arising from the composite PD/PA characteristic is given by

$$v_{im3} = \frac{3}{4}\left(a_1 b_3 - a_3\right)v_{in}^{\ 3} - \frac{25}{8}\left(3a_3 b_3\right)v_{in}^{\ 5}$$

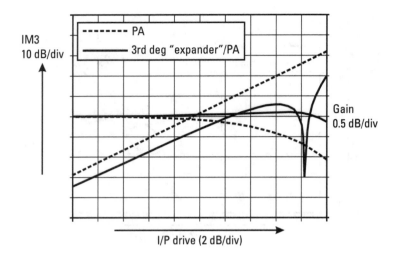

Figure 5.5 PA with third-degree nonlinearity (dotted); composite PD/PA response with PD having "unmatched" expansion characteristics (solid; $b_3 > a_3$, $b_5 = 0$).

which can vanish at a specific level of v_{in}, as shown in Figure 5.5. This desirable trait is tempered by a residual third-degree nonlinearity which reduces, but does not eliminate, the IM3 product at lower drive levels.

Unfortunately, as we will presently discover, any signal more complicated than a simple two-carrier type will have multiple third-order sidebands, and the cancellation will not occur for each IM3 sideband at equal levels of drive power. This is a warning against reading too much into predistorter results using only two-carrier tests; all too often the deep nulls such as seen in Figure 5.5 do not pan out in multicarrier or spread spectrum signal environments.

Although the above three cases have been analyzed using a PA with a very simple distortion characteristic, we will see that the results will hold up in much more generalized cases, where the PA has higher-degree nonlinearity and also AM-PM distortion. Cases 1 and 2 will be referred to as "matched" predistortion, where the successive PA nonlinear coefficients are cancelled, in theory, for all levels of input drive. Case 3 is an example of a "notching" predistorter, where cancellation takes place only at specific drive levels, in the form of nulls in the IM3 or spectral regrowth response. By and large, these two categories also distinguish different classes of practical PD realization. Simple diode PDs will usually be notchers, whereas the matched PD characteristic provides a recipe for synthesizing a more robust PD design. Separate functional analog blocks can be used to derive the different power terms, or the PD power series can be viewed as an algorithm for DSP implementation.

5.3 PD Characteristic for General PA Model

The characterization of a nonlinear PA using a Volterra series was discussed in Chapter 4. Using the techniques outlined there, it is possible to model a given PA in terms of a selected degree of power terms a_n and their corresponding phase angles, φ_n.

In almost any case, third- and fifth-degree terms will be needed, even to obtain a useful approximation to the PA compression and AM-PM characteristic, and the use of yet higher-degree terms may be necessary if the PA is being driven up to its 1-dB compression point. For the purposes of this section it will be assumed that this work has been done, and the focus is now to derive a more generalized version of the PD characteristic. Clearly, the performance of the predistorter will be highly dependent on the precision of the PA model. The analysis presented in this section represents an important step up in complexity and practical application, as compared to the ideal third-degree PA results obtained in Section 5.2. It is, however, still some way from being a completely generalized analysis of the predistorter inversion problem, to which the dedicated reader is referred [1].

Figure 5.6 defines the system and symbols used in the foregoing analysis. The analysis will show the fundamentally straightforward procedure by which a set of matched PD Volterra coefficients $[b_m, \varphi_m]$ can, up to a chosen degree, be derived as functions of a given PA characteristic, $[a_n, \phi_n]$.[2] The specific case shown here, which is one that is used in later examples, is for n, $m = 5$. The process described can clearly be extended indefinitely to handle higher-degree models for the PA than fifth; the precision which can be obtained from a fifth-degree model has been discussed in Chapter 3. We stress once again, however, that even in cases where a simple third- or fifth-

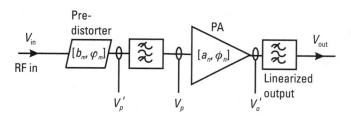

Figure 5.6 PD/PA analysis configuration.

---

2. Note that the Volterra angle symbol for the PA is now ϕ_n and the corresponding angle for the PD is φ_m.

degree model may be considered "inadequate" to replicate all of the small gyrations of PA gain over a wide dynamic range, much value can still be obtained from nulling the lower degrees of nonlinearity using a PD.

This analysis assumes that the hardware and signal environment is compatible with the envelope domain formulation, also discussed before. This allows substantial simplification in performing the inversion of the PA characteristic, in that it is only necessary to consider the AM-AM and AM-PM for a cw signal over a specified range; effects such as IM distortion and spectral regrowth are then relegated to distortion in the envelope, or baseband, frequency domain. This implies an assumption that the Volterra coefficients are invariant over a timescale corresponding to the modulation of the final signal. As discussed in Chapter 3, this assumption in effect ignores any memory effects in the RF components, which will always start to become significant at some level of precision in the whole linearization process.

The input signal to the PD, therefore, has the form

$$V_{in}(t) = V(\tau)\cos\omega t$$

where using the convention adopted in this book, a "τ" symbol indicates time in the envelope domain, which will be orders of magnitude slower than RF domain time, t. So the predistorted signal emerging from the PD will be

$$V_p{}'(t) = b_1 V(\tau)\cos\omega t + b_3\big[V(\tau)\cos(\omega t + \varphi_3)\big]^3$$
$$+ b_5\big[V(\tau)\cos(\omega t + \varphi_5)\big]^5 + b_7\big[V(\tau)\cos(\omega t + \varphi_7)\big]^7 \tag{5.4}$$

where even-degree RF domain terms have been discarded, due to the assumed band-limited nature of the system. For the same reason, it is now possible to simplify (5.4) further, by eliminating all RF domain terms except those which fall in the RF bandpass, giving (after the application of some basic trigonometric identities)

$$V_p(t) = b_1 V(\tau)\cos\omega t + \frac{3}{4} b_3 V^3(\tau)\cos(\omega t + \varphi_3)$$
$$+ \frac{5}{8} b_5 V^5(\tau)\cos(\omega t + \varphi_5) + \frac{35}{64} b_7 V^7(\tau)\cos(\omega t + \varphi_7) \tag{5.5}$$

Now for the tricky bit; we have to substitute the above expression for $V_p(t)$ into the PA characteristic,

$$V_o^{'}(t) = a_1 V_p(\omega t) + a_3 V_p^{3}(\omega t + \phi_3) + a_5 V_p^{5}(\omega t + \phi_5)$$

giving

$$V_o^{'} = a_1 \left\{ \begin{array}{l} b_1 V \cos(\omega t) \\ + \dfrac{3}{4} b_3 V^3 \cos(\omega t + \varphi_3) + \dfrac{5}{8} b_5 V^5 \cos(\omega t + \varphi_5) \\ + \dfrac{35}{64} b_7 V^7 \cos(\omega t + \varphi_7) \end{array} \right\}$$

$$+ a_3 \left\{ \begin{array}{l} b_1 V \cos(\omega t + \phi_3) + \dfrac{3}{4} b_3 V^3 \cos(\omega t + \varphi_3 + \phi_3) \\ + \dfrac{5}{8} b_5 V^5 \cos(\omega t + \varphi_5 + \phi_3) + \dfrac{35}{64} b_7 V^7 \cos(\omega t + \varphi_7 + \phi_3) \end{array} \right\}^3$$

$$+ a_5 \left\{ \begin{array}{l} b_1 V \cos(\omega t + \phi_5) + \dfrac{3}{4} b_3 V^3 \cos(\omega t + \varphi_3 + \phi_5) \\ + \dfrac{5}{8} b_5 V^5 \cos(\omega t + \varphi_5 + \phi_5) + \dfrac{35}{64} b_7 V^7 \cos(\omega t + \varphi_7 + \phi_5) \end{array} \right\}^5$$

$$(5.6)$$

It is now necessary to extract from the expansions of the above expression all terms in V^3 and V^5. This involves some detailed working, which given the help available nowadays from mathematical software will not be reproduced in detail.

The third-degree, V^3 terms are fairly easy to see,

$$\{V^3\}_{out} = \left[a_1 \left(\frac{3}{4} \right) b_3 \cos(\omega t + \varphi_3) + a_3 \left(\frac{3}{4} \right) b_1^{3} \cos(\omega t + \phi_3) \right] V^3$$

So in order to eliminate all third-degree distortion, at any level of V, this expression must be equal to zero, and making the usual normalization of the linear gain terms, $(a_1, b_1 = 1)$, the necessary third-degree predistorter coefficients are given by

$$b_3 = a_3 / a_1$$
$$\varphi_3 = \phi_3 + \pi$$

$$(5.7a)$$

the intuitive result extended to include AM-PM effects.

The fifth-degree term is

$$\{V^5\}_{out} = \begin{bmatrix} a_1\left(\dfrac{5}{8}\right)b_5\cos(\omega t + \varphi_5) + 3a_3\left(\dfrac{3}{4}\right)b_3\cos(\omega t + \phi_3) \\ b_1^{\,2}\left(b_3\cos(\omega t + \phi_3 + \varphi_3)\right)^2 \\ +a_5\left(b_1^{\,5}\right)\cos^5(\omega t + \phi_5) \end{bmatrix} V^5$$

which after some further expansion and band limiting, setting the now known results for the values of b_3 and φ_3, and making the usual linear gain normalization ($a_1 = b_1 = 1$), simplifies to

$$\{V^5\}_{out} = \left(\frac{5}{8}\right)b_5\cos(\omega t + \varphi_5) - \left(\frac{9}{8}\right)a_3^{\,2}\cos(\omega t + 2\phi_3)$$

$$- \left(\frac{9}{16}\right)a_3^{\,2}\cos(\omega t) + \left(\frac{5}{8}\right)a_5\cos(\omega t + \phi_5)$$

which as before can now be set to zero to eliminate all in-band fifth-degree distortion, and obtain relationships for b_5 and φ_5 in terms of the PA parameters. Solving the $\cos\omega t$ phasor geometry gives the relationships in the more convenient form,

$$\left(\frac{5}{8}\right)b_5\cos\varphi_5 = \left(\frac{9}{16}\right)a_3^{\,3} + \left(\frac{9}{8}\right)a_3^{\,2}\cos 2\phi_3 - \left(\frac{5}{8}\right)a_5\cos\phi_5$$

$$\left(\frac{5}{8}\right)b_5\sin\varphi_5 = \left(\frac{9}{8}\right)a_3^{\,2}\sin 2\phi_3 - \left(\frac{5}{8}\right)a_5\sin\phi_5 \tag{5.7b}$$

from which, taking appropriate care over the extraction of arctangents, the required values for the PD fifth-degree coefficients b_5 and φ_5 can be computed.

Clearly (and preferably with the assistance of a math solver), the above process can be extended to any higher order desired, either for PA, PD, or both. The key issue in this analysis is to show that the problem is completely tractable; the lower degree solutions can be extracted first, enabling higher-degree solutions to follow. This is not at all obvious in the initial formulation of the problem, such as in (5.6). So the inverted PA Volterra series, of which (5.7) is an example up to and including fifth-degree effects, is an important baseline on which to define an *a priori* method for synthesizing a predistorter characteristic for a given PA. This applies both to DSP and analog approaches.

We can now repeat the analysis of a PD/PA combination using the three cases considered in Section 5.2, except now both the PA and PD will have more realistic distortion characteristics, including higher-degree AM-PM effects. We define a PA characteristic such that $a_3 = 0.1$, $\phi_3 = 150°$, $a_5 = 0.2$, $\phi_5 = 170°$ ($a_1 = 1$), giving AM-AM and AM-PM power sweeps shown in Figure 5.7. Note that the IM3 slope is now not a straight line; at higher drive levels fifth-degree effects start to dominate and the IM3 response tends towards a 5:1 slope.

Case 1: PD matched to third-degree only

For this case, we have simply $b_3 = 0.1$, $\varphi_3 = -30°$.

Figure 5.7 shows the two-carrier IM3 response up to the 1-dB compression point, for the PA alone and the PD/PA combination. The basic picture is the same as for the simpler case considered earlier; the PD removes the 3:1 sloped IM3 component generated by the third-degree PA nonlinearity, but there is a residual IM3 response which is close to a 5:1 slope. Of course, no improvement can be expected in IM5 in this case, and is not plotted. It is of interest now, using a more realistic PA model, to investigate the sensitivity of the PD phasing term, φ_3, and Figure 5.8 shows the effect of the PD having

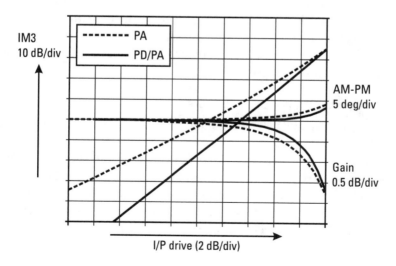

Figure 5.7 PA with third- and fifth-degree nonlinearity (dotted); composite PD/PA response with "matched" third-degree PD (solid).

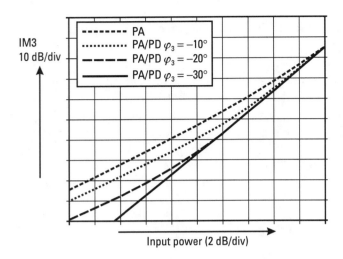

Figure 5.8 Effect of PD phase angle (φ_3) on composite PD/PA IM3.

several different values for φ_3. Here we see an immediate conflict with what may seem intuitive expectations; the "incorrect" value of φ_3 causes a substantial reduction in the IM3 cancellation *at all drive levels*, not just at the high end of the drive-power range where AM-PM effects become observable.

In a mathematical sense, this is not a surprising result; it is simply the outcome of subtracting two vectors having the same magnitude but non-aligned directions. Unlike AM-PM distortion at the fundamental, where the relative strength of the linear term swamps the effect of the third-degree phase differential at low drive levels, the IM3 signal has a steady φ_3 phase offset from the input at all (lower) drive levels.

Case 2: PD matched, third- and fifth-degree

Using the results in (5.7), we obtain $b_3 = 0.1$, $\varphi_3 = -30°$, $b_5 = 0.221$, $\varphi_5 = -13$.

Figure 5.9 shows IM3 plots for the PA alone and the PD/PA combination. As expected, there is now a residual 7:1 slope on the IM3, showing that third- and fifth-degree distortion has been successfully removed, but a seventh-degree distortion "residue" has been left behind by the predistortion process. The IM5 plot will also show a 7:1 characteristic. Both of the PD results will show the same sensitivity to the PD phase angles, as discussed in case 1 above.

This is an ideal result, assuming not only that the PD parameters can be precisely realized in practice, but that the PA itself can be modeled over

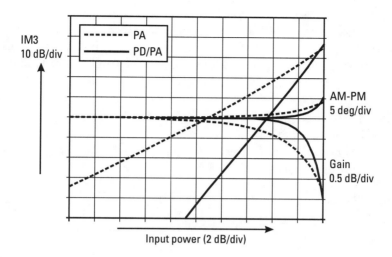

Figure 5.9 PA with third- and fifth-degree nonlinearity (dotted); composite PD/PA response with matched third- and fifth-degree PD (solid).

the required dynamic input signal range using a fifth-degree Volterra series. But it sets a target and compiles a recipe for practical predistorter design.

Case 3: "Unmatched PD"

Clearly, there is wide scope for illustrating the effects of PDs which have unmatched, or not perfectly matched, characteristics to the PA. Such cases result in notched IM responses, such as shown already in Figure 5.5. As discussed previously, IM notches at specific power levels are still possible, although inclusion of AM-PM effects in the PA and PD characteristics tends to reduce the depth of the notches in IM power sweeps. There is also a gray area between cases 2 and 3, where the PD characteristics are close but not precisely equal to the ideal values. Such cases represent many practical situations.

The above cases have been analyzed using a simple two-carrier signal environment. Before drawing too many conclusions about the results and their implications for PD design, it is necessary to evaluate PD performance using a more complex multicarrier signal. In order to do this, it is necessary to use a computational technique to generate the signal, impose the PD and PA characteristic successively, and perform the necessary spectral processing on the output signal. There are many commercial CAD tools available which can perform this task, and the details of the computations used to obtain the results in this section will not be given. A bigger problem in presenting such results is the large number of signal environments which are encountered in modern communications systems. It seems appropriate in the present discussion to illustrate the PD performance using a multicarrier signal. Multicarrier signals, in general, represent the biggest challenge to the PA designer. They have both high peak-to-average ratio and zero crossing, and the spectral distortion is a stronger function of the AM caused by the multicarrier effect, rather than any AM present on individual carriers.

So we now take another pass on the three cases, looking at the output spectrum for a few levels of drive power.

Case 1: PD matched to third-degree only

PA: $a_3 = 0.1, \phi_3 = 150°, a_5 = 0.2, \phi_3 = 170°$

PD: $b_3 = 0.1, \varphi_3 = -30°$

Figure 5.10 shows two 16-carrier spectral sweeps, in which successive sweeps are shifted in order to give a visible overlay effect. The first sweep shows the PA by itself, operating at a power level close to the composite 1-dB compression point; this mean drive level will now, of course, be backed off (by nearly 10 dB) from the corresponding compression level for the previous two-carrier signal. The overlaid second sweep shows the signal backed off by a further 6 dB. This enables the third- and fifth-order IM products to be distinguished, but it is notable that this division is not as well defined as in a

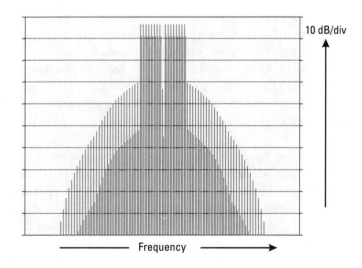

Figure 5.10 Multicarrier PA spectral response; third- and fifth-degree PA distortion (1-dB compression level and 6-dB PBO).

simple two-carrier case. The close-in third-order products have a substantial fifth-degree component, and therefore show a backoff slope somewhere between 3:1 and 5:1.

Figure 5.11 shows the same sweeps taken with a third-degree matched PD in place, with the unpredistorted PA sweeps included for direct comparison. As would be expected from the previous two-carrier results, the improvement at the 1-dB compression point is quite small, and the fifth-order products show a significant increase due to the third-degree PA characteristic operating on the third-degree predistorted input signal. The picture greatly improves as the power level is backed off; the benefits of matched third-degree predistortion become more evident at lower power levels, where third-degree effects start to dominate.

Case 2: PD matched, third- and fifth-degree

PA: $a_3 = 0.1, \phi_3 = 150°, a_5 = 0.2, \phi_3 = 150°$

PD: $b_3 = 0.1, \varphi_3 = -30°: b_5 = 0.221, \varphi_3 = -13°$

Figure 5.12 shows the same power/spectrum sweep conditions as in Figure 5.11, the difference being the inclusion of the fifth-degree matched term in the PD characteristic.

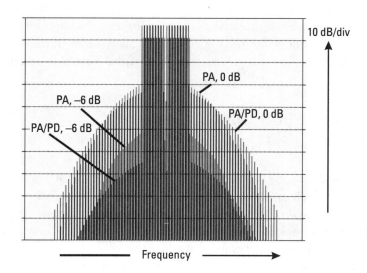

Figure 5.11 Multicarrier PD/PA spectral response; third-degree (only) PD, 1-dB compression point and 6-dB PBO. Original PA responses (from Figure 5.10) are also shown for comparison.

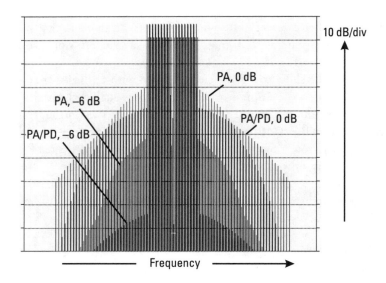

Figure 5.12 Multicarrier PD/PA spectral response; "matched" third- and fifth-degree PD.

The higher (1-dB compression) drive level shows a substantial improvement in the suppression of close-in third-order products, although at the expense of some increase in the level of the more extreme fifth-order products. These increases are due to the generation of seventh-degree effects by the fifth-degree PD, and will show a 7:1 backoff. This is shown in the 6-dB backoff sweeps, which show major reductions in all visible distortion products.

Case 3: A "notcher"

PA: $a_3 = 0.1, \phi_3 = 0°, a_5 = 0.2, \phi_3 = 0°$

PD: $b_3 = 0.2, \varphi_3 = 0: b_5 = 0, \varphi_3 = 0$

Figure 5.13 shows the important difference between the matched PD and the notcher, which causes cancellation at a single drive level. In a multicarrier environment, the notcher looks much less attractive, in that the cancellation only occurs at a single IM offset frequency, as well as a single power level.

The last result in Section 5.2 confirms an observation which is made frequently by researchers in this field; notching effects are much easier to demonstrate under two-carrier test conditions, and can "fill-up" in multicarrier or even single-carrier spread spectrum QPSK environments. This section presents a quantitative analysis which indicates how and why this is so.

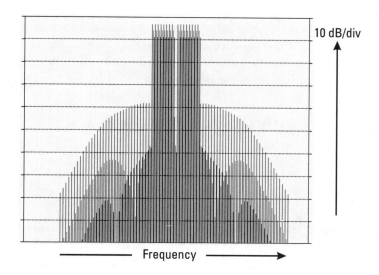

10 dB/div

Frequency

Figure 5.13 Multicarrier PD/PA spectral response; "notcher" PD. PD/PA responses only for PBO levels –3 dB, –6 dB, and –9 dB.

We assume that the final PD/PA characteristic has the form

$$v_o = a_1 v + a_3 v^3 + a_5 v^5 + a_7 v^7 \dots \tag{5.8}$$

where "v" is the input excitation, having the form

$$v = V(\tau)\cos(\omega t + \Phi(\tau)) \tag{5.9}$$

where τ represents the "slow time" of the modulation domain. For simplicity, and without any loss of validity in demonstrating the issues under consideration in this section, the phase terms in the PD/PA characteristic will be ignored. Note also that the PD/PA response, (5.8), is redefined now in terms of a "composite" amplifier, with its own set of a_n coefficients. These a_n coefficients incorporate the effects of an internal predistorter, but could also in some cases represent an amplifier which has not been designed specifically with a predistorter; notching effects of the kind under discussion here can also be observed in some types of standalone PAs, as discussed in Chapter 1.

In the simplest case of a two-carrier signal, (5.9) becomes

$$v = V\cos(\Omega t)\cos(\omega t)$$

where V is now a scalar amplitude and the modulation domain time, τ, is expressed in terms of the modulating frequency, Ω, so that just as $\tau \gg t$, so $\Omega \ll \omega$. The two-carrier situation was already analyzed in Section 5.2, so that removing the Volterra phase angles from (5.5) gives, for the present case,

$$v_o(t) = \begin{bmatrix} a_1 V \cos(\Omega t) + \dfrac{3}{4} a_3 V^3 \cos(\Omega t) + \dfrac{5}{8} a_5 V^5 \cos^5(\Omega t) \\[2mm] + \dfrac{35}{64} a_7 V^7 \cos^7(\Omega t) \end{bmatrix} \cos(\omega t)$$

(5.10)

The IM3 products come from the third harmonic component of $\cos \Omega t$, modulating the RF carrier $\cos \omega t$. All higher-degree $\cos(\Omega t)$ terms in (5.10) will therefore generate a contribution to the final IM3 value. In accordance with a fifth-degree truncation, the IM3 products can be written as the sum of these third- and fifth-degree components.

In order to underline the key issue in this analysis, the cognizant trigonometric identities are noted at this point:

$$\cos^3 \theta = \frac{1}{4}(3\cos\theta + \cos 3\theta)$$

$$\cos^5 \theta = \frac{1}{16}(10\cos\theta + 5\cos 3\theta + \cos 5\theta)$$

$$\cos\alpha\cos\beta = \frac{1}{2}\big[\cos(\alpha+\beta) + \cos(\alpha-\beta)\big]$$

so that the IM3 amplitude is

$$V_{im3} = a_3 V^3 \left[\frac{1}{2}\frac{3}{4}\frac{1}{4}\right] + a_5 V^5 \left[\frac{1}{2}\frac{5}{8}\frac{5}{16}\right]$$

(5.11)

and this will be the amplitude of the upper and lower IM3 sideband, AM-PM effects having been ignored for this analysis.

As noted previously, the IM3 amplitude (5.11) can be made to vanish in two distinct ways. Clearly, in the case of a well-designed, matched predistorter, the coefficients a_3 and a_5 will be zero, so the IM3 "null" will have a wide dynamic range, essentially independent on the signal amplitude V. In what represents a wide range of practical situations, which can even include attempts to obtain matched PD performance, the a_3 and a_5 coefficients will still have finite values, which may be opposed in sign, so that cancellation occurs at a single, specific value of V. What distinguishes the two-carrier case is that the IM3 products only occur at a single offset frequency, Ω. Anything more complicated in terms of a signal envelope function, $V(\tau)$, will result in multiple IM3, and the same for higher order (e.g., IM5 and IM7) sidebands. For example, consider the simplest next step in signal complexity, a four-carrier system, which can be represented as

$$v = V\{\cos(\Omega t) + \cos(2\Omega t)\}\cos(\omega t) \tag{5.12}$$

This can be easily seen to be a specific case of four RF carriers, with two pairs spaced Ω apart, and a central 2-Ω spacing between the pairs. The in-band IM3 products generated by a third-degree nonlinearity will have the form

$$V_{im3} = \frac{3}{4} a_3 V^3 \{\cos(\Omega t) + \cos(2\Omega t)\}^3 \cos(\omega t)$$

which can be expanded to show now a substantial band of IM3 products in the envelope frequency domain,

$$\left[V_{im3}\right]_{3rd\ deg} = \frac{3}{16} a_3 V^3 \begin{Bmatrix} 3 + 9\cos\Omega t + 9\cos 2\Omega t + 4\cos 3\Omega t + \\ 3\cos 4\Omega t + 3\cos 5\Omega t + \cos 6\Omega t \end{Bmatrix} \cos\omega t \tag{5.13}$$

as shown in Figure 5.14.

The key issue here is that the amplitudes of the multiple sidebands vary considerably over the IM3 band, most notably the extreme sidebands having substantially lower amplitude than the inner ones. Note also that the highest magnitude IM3 products, those at the $\omega \pm \Omega, \omega \pm 2\Omega$ points are essentially now gain compression components since they lie underneath original input carriers; the third-order "spectral regrowth" band comprises the frequency range $\omega \pm 2\Omega$ out to $\omega \pm 6\Omega$, which shows a 4:1 voltage range, or 12-dB power range, in the sideband amplitude.

A notching predistorter will typically show notching behavior with this type of signal, but the notches will occur at different drive levels for the different sidebands in the IM3 spectral regrowth frequency band. With considerably greater demands on algebraic motivation, the four-carrier signal function (5.12) can be substituted into the fifth-degree term of the PA/PD

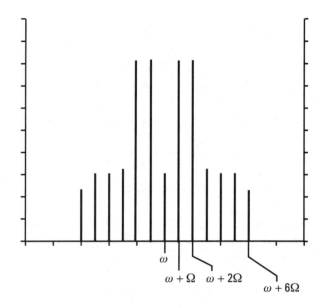

Figure 5.14 Four-carrier third-order IM spectrum.

characteristic, and fifth-degree contributions to the sidebands in the third-order spectral regrowth band, given in (5.13), can be evaluated.

The results of this are

$$[V_{im3}]_{5th\,deg} = \frac{5}{8}\frac{1}{16}a_5 V^5 \left\{ \begin{array}{l} 30 + 110\cos\Omega t + 105\cos 2\Omega t \\ +75\cos 3\Omega t + 60\cos 4\Omega t \\ +51\cos 5\Omega t + 30\cos 6\Omega t \end{array} \right\} \cos\omega t$$

$$(5.14)$$

from which it can be seen that the nulls in the separate IM3 sidebands, which can be obtained by summing the individual contributions from (5.13) and (5.14) and setting to zero, will occur at different levels of V, there being no duplications in the various sideband coefficients in the fifth-degree expansion.

This is just a simple step towards a signal which has many more carriers, or a spread spectrum signal. In each case, the essential mathematical behavior analyzed here will be observed; the notching effect of a simple predistorter can be seen to "travel" along the spectral regrowth zone as the signal amplitude is varied.

5.4 Practical Realization of the Predistorter Function: Introduction

The results in Section 5.3 show that a predistortion function can be derived as a mathematical function for a PA with a given characteristic. In practice, of course, the characteristic of a given PA may not be accurately described by a fifth-degree Volterra series, even if phase angles are included. This will, unfortunately, usually be more apparent as efficiency enhancement tricks are employed; the tradeoff between efficiency and linearity is one which nature seems to hold particularly close. It is, nevertheless a fact that even if higher-degree nonlinearities are playing a significant role in the upper end of the power range, major reductions in IM and spectral distortion can be achieved with the appropriate third- and fifth-degree correction. A more troublesome issue is the variability of the measured PA parameters with different signal environments. This subject was discussed in Chapter 3, and raises the issue, indeed in some applications the necessity, of having a predistortion scheme which can be adapted to varying signal conditions. In describing various practical approaches to realizing the PD function, therefore, it is important to recognize the possible need to adapt, or modify, the PD parameters using external control signals.

It is clear that in the modern era of high-speed DSP, if a PD distortion function can be defined algorithmically, the most obvious method for its implementation is to use DSP hardware and appropriate software algorithms. It is, however, still very relevant to consider possible implementation schemes using analog hardware. The advantages and disadvantages of each approach will be discussed in parallel with their descriptions in the following section, but the overall conclusion is that a combined approach may be the best way of leveraging the main benefits of both methods: speed and simplicity for analog, and algorithmic precision for DSP.

Whether DSP or analog in nature, there are two basic methods by which the input signal to a PA may be predistorted. These are shown, in their simplest schematic form, in Figure 5.15. The first method, Figure 5.15(a), could be described as the traditional predistorter, whereby a physical nonlinear device is used. This device, usually consisting of one or several diodes, has to be tailored to have the best possible approximation to the required PD characteristic. The alternative approach is shown in Figure 5.15(b). Here the PA has a gain and phase modulator placed on its input, so that the gain and phase of the PA can be adjusted in accordance with its previously measured nonlinear properties. The modulator therefore requires a two-dimensional drive signal to perform its task.

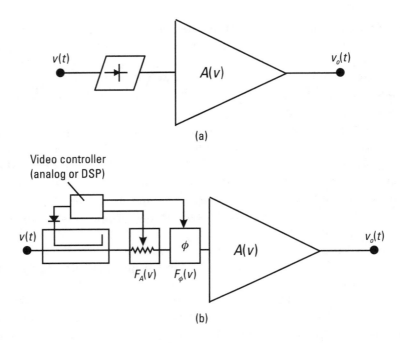

Figure 5.15 Predistortion categories: (a) nonlinear element and (b) vector modulator.

The process by which these drive signals are generated, in a dynamic signal environment, opens up several sub-categories of PD architecture, which have formed the basis for numerous proprietary products and patents over the years.[3] Most make the basic assumption that the gain compression and AM-PM are functions of the "current" envelope amplitude. As discussed at some length in Chapter 3, this assumption is quite approximate, and ignores memory effects, whereby both the amplitude and phase distortion may have additional dependency on immediately past values of the envelope amplitude. Although a PD based on such an assumption may give useful performance for some applications, memory effects constitute an ultimate limitation in the performance of an open-loop linearization scheme. The DSP approach does, of course, include the possibility for implementing a suitable algorithm for memory effects in addition to the basic predistortion function.

---

3. The reader is referred to the preface of this book for a general statement on patent issues; this is a heavily patented area and any commercial implementation of the techniques being described should not be undertaken without thorough patent searches and appropriate actions thereon.

5.5 Analog Predistorters

Analog predistorters come in two important sub-categories. The "simple" predistorter, which usually consists of a configuration of one or more diodes, and the "compound" predistorter, which can in principle synthesize the required nonlinear characteristic using separate sections to generate the various degrees of distortion. The simple PD relies fundamentally on selecting and/or tailoring the nonlinear characteristics of the PD device to match, or cancel, the PA nonlinearities; the compound PD does not rely on the nonlinear elements having specifically tailored characteristics. The compound PD is a less familiar concept and will be the main focus in this section.

Simple analog PD circuits abound in the literature [2–4]. They mainly use a nonlinear resistive element such as a diode or an FET channel as an RF voltage-dependent resistor, which can be configured to provide higher attenuation at low drive levels and lower attenuation at high drive levels. This principle is illustrated in Figure 5.16, along with some attenuation characteristics which show a tradeoff between low-level insertion loss and the useful range of gain expansion. In constructing simple predistortion circuits of this kind, it is important to provide a path for the rectified dc in a manner that does not slow down the response. One of the main advantages of these simple PD circuits is that they work "on-the-fly" and can handle signal envelopes which vary at speeds up to less than an order of magnitude down from

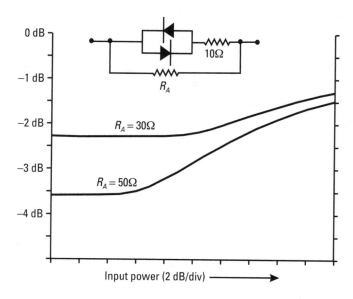

Figure 5.16 Simple series diode gain expander.

the RF cycle time. Consequently, a more complicated configuration such as that shown in Figure 5.17 represents a different tradeoff between simplicity, performance and signal bandwidth. Here a PIN diode is used as the controlling element and is driven by a detector which senses the signal level. The necessary scaling and offsetting of the detector output to form the necessary drive to the PIN diode will usually involve the use of an operational amplifier, and consequently several orders of magnitude of speed are sacrificed.

The typical performance of simple PDs of this type will show much similarity to the "unmatched" predistorters analyzed in previous sections. It is usually quite easy to adjust the PD settings such that the gain compression of the PA is cancelled somewhere in the vicinity of the 1-dB compression point. This will result in a sharp null in the IM characteristics in a simple two-carrier test at about the same RF drive level as the cw cancellation. As discussed in Section 5.3, however, the deep nulls tend to fill in when the device is tested using a more complex signal. But these simple PD devices are still useful, even valuable, in some applications such as RFIC PAs for battery-operated handsets. One of the main factors which has limited their use in these applications is the critical nature of the adjustments required to position the null points. Such adjustments are undesirable in high-volume applications, and the limitations and hazards of placing too much reliance on an open-loop technique are apparent.

It is ironic that simple PD circuits of this kind may have been wrongly positioned in the whole predistortion scenario. It seems that one strength they may have, and which was analyzed in Section 5.3, is that for well backed-off PAs, they can provide a precisely matched third-degree characteristic which could alleviate much of the precision which would be required from a DSP controller at these levels. One of the reasons that this valuable property may have been overlooked is the importance of matching the correct Volterra phase angle; even a 10° error can virtually eliminate any low-level correction (see Figure 5.8). The phase performance of simple PDs tends

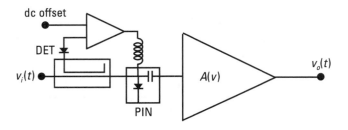

Figure 5.17 PIN diode attenuator driven by input peak detector.

to be given only secondary consideration. In general, the accompanying AM-PM predistorter characteristic can be tailored by varying a series or shunt reactive element, but above about 1 GHz such a design process will be severely limited by the package reactances of surface-mount (SMT) components. Once again, the RFIC designer has much greater scope in this area.

One interesting possibility for an RFIC predistorter is to use the saturation characteristic of a mesa resistor, rather than a diode, as the nonlinear element. The characteristic of a mesa resistor in a typical GaAs process is shown in Figure 5.18. It has the general appearance of a saturating MESFET I-V characteristic, but without the gate control. Such an element, placed in shunt with a 50-Ohm transmission line, will show low resistance at low RF signal levels (the normal regime for linear resistors using the MMIC process) which will transition to a much higher resistance as the drive signal swings the voltage into the saturation region. The drive level at which the transition occurs can be set by suitable choice of resistor dimensions, and the phase shift (AM-PM) can also be set using a shunt capacitance.

All of the above predistortion devices have substantial limitations. What is really needed is a method by which a given PD characteristic can be "synthesized," preferably using nonlinear devices whose properties do not have to be precisely crafted in each individual case. An important concept in making this critical step is shown in Figure 5.19. An incoming signal is split into two paths, and recombined with a 180° phase shift at the output. One path contains a nonlinear element, the other path contains a variable attenuator and delay line which can be adjusted to cancel the linear components emerging from each path. The output signal emerging from the combiner, assuming a band-limited situation, now contains components proportional to the third-, fifth-, and so forth degree powers of the input signal. In mathematical terms, if the input signal is $v(t)$, then the upper path will produce a signal

$$v_u(t) = b_1 v(t + \tau_1) + b_3 v^3(t + \tau_3) + b_5 v^5(t + \tau_5) + \dots \quad (5.15)$$

and the lower path will give a second signal,

$$v_u(t) = b_1 v(t + \tau_1)$$

so the output 180° combiner (balun) will form the difference between these two inputs, giving an output signal

$$v_u(t) = -b_3 v^3(t + \tau_3) - b_5 v^5(t + \tau_5) + \dots \quad (5.16)$$

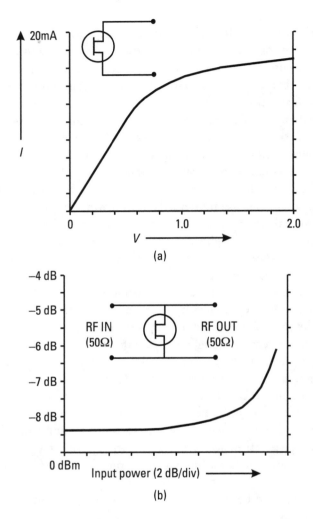

Figure 5.18 Mesa resistor as predistorter: (a) I-V characteristics of "gateless FET" and (b) RF characteristics using packaged low noise FET (NE760).

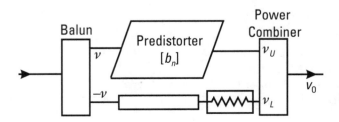

Figure 5.19 Basic "cuber" configuration.

The key point about this signal is that the distortion terms can now be scaled and phase shifted independently from the original undistorted input signal. They cannot, of course, be so scaled with respect to each other (i.e., third and fifth), but for practical purposes it may be possible to set the drive level such that only the third-degree term is significant, so that at least the third-degree distortion term can be isolated. For these reasons, this device is sometimes called, generically, a "cuber." Specific analog implementations based on this concept were quite widely implemented, and usually patented, to generate polynomial-based predistortion functions in the pre-DSP era [5]. It bears some resemblance to the first loop in a feedforward system, but in practice is a much simpler piece of hardware. In particular, the first-order cancellation process does not need to be nearly so precise as that required in a useful feedforward system.

Using a cuber of this kind, it is in principle possible to construct an ideal, matched third-degree predistorter as discussed in Section 5.3. As shown in Figure 5.7, such a device has useful and robust characteristics, only falling short in the compression region where some higher-degree correction would be desirable. A basic schematic is shown in Figure 5.20. The input signal is split, one part being the main signal line to the PA input, and the other forming the input to the cuber. The cuber output is recombined with the input signal to the main PA, following amplitude scaling and phase-shifting elements. The key point about this predistortion scheme is that the amplitude and phase controls at the cuber output can be used to *set* the effective third-degree predistorter coefficients. This can be done using almost any kind of nonlinear element in the cuber; the need to tailor a prescribed nonlinear device has been eliminated.

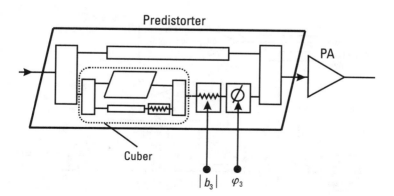

Figure 5.20 Cuber used as predistorter.

This valuable property of compound predistortion is worthy of practical illustration. Figure 5.21 shows a schematic of a simple back-to-back diode limiter, and a hardware realization using an FR4 test board and a pair of SMT Schottky diodes (5082-2810). The measured swept power-limiting characteristic is shown in Figure 5.22. Clearly, such a device has a compression characteristic and is not a candidate, in this configuration, for use as a simple predistorter which would require gain expansion. Both the compression and the AM-PM sweeps do, however, show quite characteristic PA behavior. This similarity can be harnessed by placing the device into a cuber cancellation circuit, as illustrated schematically in Figure 5.20. The first-degree cancellation can be readily achieved on a test bench using a pair of equal-power splitters, along with suitable gain and phase trimmers. The

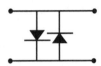

Figure 5.21 Simple diode limiter.

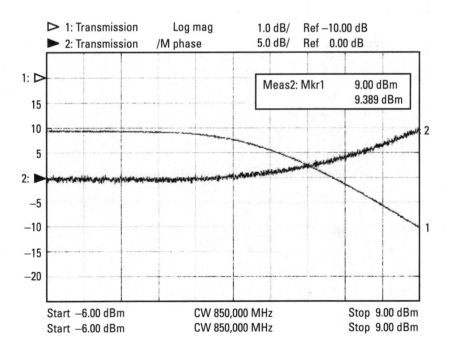

Figure 5.22 Measured transmission characteristics of shunt diode limiter.

resulting power sweep response is shown in Figure 5.23. The key observation is the smooth 2:1 "gain" slope, indicating a well-behaved third-degree nonlinearity in the backed-off region, up to about 6 dB backed off from the 1-dB "compression" point of the limiter. The corresponding phase angle can be seen to be a steady value in this region, about 135°. This output signal from the cuber can now be scaled and phase-shifted to give a precise third-degree correction to any amplifier.

It should be pointed out that there are some practical inconveniences in this otherwise promising contrivance. The combiners and splitters will create a device with a significant insertion loss in the input to the PA. There is a tradeoff; lower coupling factors into and out of the cuber result in lower main path attenuation, but the nonlinear element will need to have the required performance at a lower drive-power range. Equal-power splitters and combiners will give a minimum of 6-dB insertion loss in the main line, which with typical losses may grow to 7 or 8 dB. This is hardly a problem for a high-power PA with many stages, but would be an issue in a low-cost on-chip implementation. Temperature and aging effects also have to be considered.

There remains an issue concerning the higher-degree nonlinearities which appear at the output of the cuber and which will not, in general, be matched to the requirements of the PA. One possible approach to this issue is to consider the use of a nonlinear element in the cuber which has a

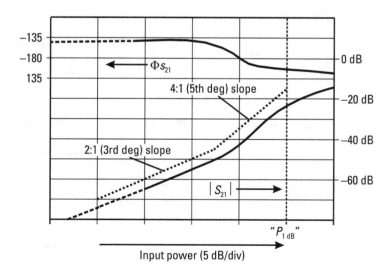

Figure 5.23 Measured performance of cuber, using a simple diode limiter as a nonlinear element. (Dotted lines indicate data below system noise level.)

characteristic that matches the PA. This is a seemingly simpler task than try-ing to create a device with an inverted PA nonlinearity. Unfortunately, recall-ing the results in (5.7a) and (5.7b), a simple phase inversion, or "negation," of the fifth-degree PA nonlinear coefficient is not the correct fifth-order PD coefficient.[4] So the use of a smaller periphery, or a small section of the PA output transistor, as the cuber nonlinear element is not, theoretically, an acceptable solution.

Figure 5.24 shows a possible method by which both third- and fifth-degree nonlinear terms can be separated. The signal is now split between two separate cubers, one being set through an input adjustment (or asymmetrical input coupling factors) at a substantially higher drive level such that it gen-erates much higher fifth-degree distortion than the other. Then by suitable scaling and recombining it is possible to cancel the third-degree distortion signals at one cuber output, and cancel the fifth-degree signals at the other. Each distortion signal can then be scaled and phase-shifted. Such a device has been proposed in principle, but not described in detail [6]. Although it may appear that DSP now offers a more logical approach to implementing algo-rithmic nonlinear functions, the inherent speed of the analog approach may still have a part to play.

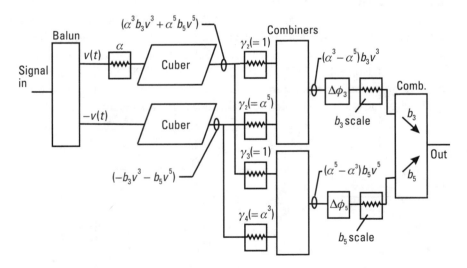

Figure 5.24 Possible configuration for generation of third- and fifth-degree predistortion signals, with independent controls on magnitude and phase angle.

---

4. This is unlike the third-degree term, which is a simple negation.

One of the difficulties in a practical realization of the system shown in Figure 5.24 would appear to be the longevity of the various attenuation and phase settings over changing signal and environmental conditions. It seems all too easy to dismiss the concept as a tweaker's nightmare, having a useful operating lifetime measured in minutes. But the same could, and will, be said of a feedforward loop in a basic manually adjusted implementation. Both systems need the care and attention of an adaptive software monitoring system, and it seems only fair to consider the various different linearization approaches on equal terms. Another issue is the relative complexity of the hardware, which essentially attempts to perform analog computations on the incoming signal. Such computations could, in principle, be performed with ease by DSP, but again a fair comparison has to be made. The compound predistorter works on a sample of the original signal and delivers a suitably distorted version directly to the PA. The process is fast, and achieving signal bandwidths up to even 10% of the RF carrier would be quite feasible using carefully designed RF components. A DSP system attempting the same task first has to translate the signal into a form where the computations can be performed, and then reconstruct, or modify, the signal to form the final output. With the availability of high-speed DSP hardware, the digital approach is quickly becoming the favored approach, but it is important to recognize that the underlying predistortion task remains the same.

The compound cuber would appear to emerge from the above discussion as a method for obtaining a robust linearization performance using a piece of microwave hardware which consumes little or no power. Its strength lies in an ability to generate a precisely scaled and phased third-degree correction signal at backed-off drive levels where third-degree effects in the PA dominate the generation of close-to-carrier distortion. A corresponding improvement in EVM will also be obtained. Such devices represent an important and major step forward in predistorter design, from the simple traditional diode expander. There may be some benefits from using such analog predistortion configurations with DSP adaptation, as compared to the conventional methods of applying DSP correction.

5.6 DSP Predistortion

Referring to Figure 5.15, the use of an input gain and phase modulator is a quite distinct method of predistorting a signal, and it is justifiable to ask whether the detailed mathematical formulation of PD characteristics given in previous sections is still applicable. Basically, the modulator synthesizes the entire PD response, to however many degrees are used, at each level of the applied RF signal; the Volterra characteristic is turned back into a gain-phase transfer versus RF drive level characteristic. So one method of deriving the modulator drive signals is to evaluate the required gain and phase adjustments using the predetermined PD response as an algorithm. Then, of course, it is necessary to transform these numbers into corresponding drive signals to the modulator, which may well have a Cartesian, rather than a straight polar, drive input. This second part of the process will almost certainly involve the use of a look-up table (LUT) based on *a priori* characterization of the modulator, and the question arises as to whether this kind of predistorter should not simply use an LUT in the first place, which includes both the PA predistortion requirements and the modulator drive as a single composite table entry for an appropriate density of RF drive levels. Such linearization schemes, based on LUTs, have been described extensively in the literature, and have formed the basis for some commercial linearized PA products. Some of the issues surrounding the use, and compilation, of LUTs will be discussed in this section, although the reader is referred to an extensive literature for a more detailed treatment of the DSP aspects of the subject [7–9].

Without here resorting to the mathematics, which essentially confirms the equivalence of the two predistortion methods of Figure 5.15, it is instructive to look at some simple numbers. Clearly, if the PA is running at 1-dB gain compression and 10° of AM-PM, it is possible to imagine sending appropriate signals to the modulator such that its insertion loss is reduced by somewhat more than 1 dB (to allow for further compression, as discussed in previous sections), and also to introduce an input phase shift to allow for the now considerably higher than 10° of AM-PM that the PA will display as a result of the increased input amplitude. Clearly, and as discussed in the early sections of this chapter, there will be a "point of no return" where the escalating compression of the PA will not allow any increased level to restore the output to its appropriate "linear" level. But at levels backed off by a few decibels, the predistortion modulation process quickly becomes less troublesome, the correction levels of amplitude and phase being very close to the gain compression and AM-PM values of PA performance.

At still lower levels of PBO, a different kind of problem emerges. The corrections, measured in decibels and degrees, become very small numbers and the precision required from the control signals becomes correspondingly greater. This problem is illustrated in Figure 5.2, for a simple third-degree PA nonlinearity. At the 10-dB backoff point, the compression is 0.1 dB, and at the 20-dB backoff point the compression is 0.01 dB. So in order for the linearization process to be effective in the 10–20-dB backoff range, precision measured in terms of 0.001 dB will be required. If the amplitude modulator is assumed to have a simple logarithmic attenuation drive characteristic, say, a range of 5 dB over 0–5-V drive, a precision of 0.001 dB would require a 14-bit digital-to-analog converter (DAC). Such a device is available, but has a speed limitation. There is also the issue of whether the devices used in the modulator itself will maintain operation to this degree of precision.

It should be noted that there are other applications, for example, the calibration of detector diodes in precision microwave power meters, where the low-level performance is sufficiently close to expectations based on device physics that a simple analog device may find the required precision less taxing than one using DSP controls. In other words, a simple third-degree compound cuber (as discussed in Section 5.5) having third-degree amplitude and phase coefficients well matched to those of the PA may do as good or better job than a DSP-driven modulator in this well backed-off region. This aspect of linearization probably receives less attention due to the fact that at such levels the PA distortion and EVM may well fall within specified levels in any case. But it raises the possibility of a hybrid approach, using a compound analog predistorter as the DSP control elements, rather than a simple vector modulator. Such an arrangement would have more built-in predistortion action, particularly at the lower drive levels. At higher levels, DSP optimization routines could be used to set the scaling and phasing elements to suit the changing signal and physical environment.

A more conventional approach, however, is to use an LUT to drive an input modulator, as shown in Figure 5.25. It should be noted that the use of an input signal delay can in principle, compensate for the processing delays in the detection and DSP and in this sense the system is not limited by the speed of the DSP itself. Such a device will, however, have all of the same limitations as any other form of predistortion device. In particular, the

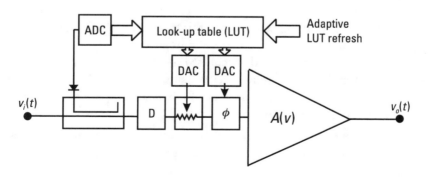

Figure 5.25 Basic DSP predistortion scheme.

escalating AM-PM characteristics of most PA devices as they enter the compression region, along with the mathematical amplitude limitation discussed earlier, will restrict the practical range of linearization. As has been discussed several times in this chapter, the correction signals have to contain multiple harmonics of the baseband signal in order to perform the necessary predistortion function; this places a more stringent requirement on the data converters than that required to generate the original signal. Having said this, it seems that up to the 1-dB compression point, a well-designed predistortion scheme should be able to reduce ACP and EVM substantially, compared to the uncorrected PA. There are, however, a number of additional practical limitations which make this a difficult performance level to achieve in practice. At the risk of a little repetition in some cases, these limitations are now listed.

1. *LUT precision.* This is a complicated issue, and a detailed treatment is not attempted in this book, but several points are worth making:

   (a) The LUT can be either physical, or implied in the form of a suitable algorithm. The final drive signals to the physical PD control lines will have to incorporate the characteristics of the drive elements (e.g., vector modulator) as well as the PA itself.

   (b) The LUT, or computational workload, will be much lower in a system which "modifies" the incoming signal, rather than one which "reconstructs" the signal. The reconstruction approach becomes more viable in situations where the complete transmitter design is being undertaken, and the predistortion can be done as part of the baseband processing.

   (c) The predistorted signal emerging from an ideal predistorter will necessarily contain harmonics of the baseband modulation signal. In practical terms, this means that the DSP drivers have to work at a speed corresponding to maybe an order of magnitude faster than that required to generate the original signal; in multiplexed multicarrier systems, the required frequency components may be an order of magnitude higher than the maximum carrier spacing.

   (d) A simple static power sweep, measuring gain compression and AM-PM will typically be neither precise enough, or even representative, of dynamic signal conditions. This issue (see Chapter 3) may well be a fundamental limit to the effectiveness of predistortion as a standalone (as opposed to a complementary) linearization technique.

(e) The effects of any changes in input and output mismatch can be quite significant on the LUT values. The final PA assembly must be well isolated from external mismatches, and any internal switching required to perform LUT refreshes must be similarly isolated in order to maintain a constant impedance environment for the PA, PD, and all of the associated monitoring circuitry.

2. *LUT longevity.* The issue of longevity often triggers a discussion on long-term PA drift, temperature effects, and related environmental issues. In fact, the longevity problem in any predistortion configuration will typically be a much shorter-term effect caused by changing signal conditions. In any Class AB amplifier, the thermal dissipation in a device varies with drive level, and the PA design should take account of this as an additional design issue. The variations will start to show themselves on different timescales as the signal environment changes. Experimental data presented in Chapter 3 shows that even at envelope speeds of 10 kHz, some hysteresis and asymmetry can be observed in a Class AB device. But at longer timescales, the time-averaged power dissipation can cause significant changes in gain and phase which may show up as additional ACP. Such changes can only be accommodated by having a dynamic LUT refreshing system. In principle, the thermal effects have a defined physical origin, and it would seem that the integration of dynamic thermal behavior into the familiar electrical nonlinear models will be required to establish an algorithmic formulation of these changes.

3. *Envelope input sensing.* It is commonly assumed that the dynamic sensing of the input signal envelope is a trivial task compared to the generation of a suitably predistorted version of it. In fact, as signal throughputs continue to rise, this can become a substantial problem in its own right. The classical envelope detector has a tradeoff between the precision of the detection process and the number of RF cycles used to determine the final detector output. This tradeoff becomes more critical as the signal bandwidth and RF carrier frequency become closer. A multicarrier signal, spread over, say, 10 MHz at a carrier frequency around 1 GHz represents

a challenge to the detector designer. The envelope amplitude measurement system, along with the necessary post-detection amplifier, will require calibration. This can in principle be integrated into the main PA LUT. Care must be taken in using any detector that thermal effects do not introduce hysteresis into the measurement at the high end of the power range.

Taking due account of the above issues, one approach to the calibration and maintenance of an LUT-based predistortion system looks quite attractive, and has already been introduced briefly in Chapter 4. Figure 5.26 shows a configuration which allows a PA to be switched between two linearization modes, analog vector envelope feedback and DSP predistortion. The gain and phase modulator elements are common to both linearization loops, so that the DSP can dynamically monitor, and save, the correction drive signals for a selected signal environment while the feedback performs the linearization in a desirable closed loop fashion. The key issue here is that the DSP LUT can be loaded using a dynamic calibration signal which is slow enough that the loop delays have a negligible effect on the linearization fidelity. This

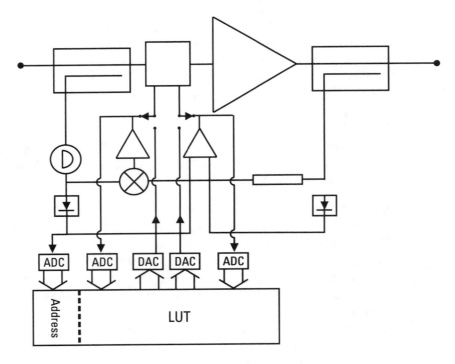

Figure 5.26 Envelope feedback used as basis for LUT calibration.

form of calibration, which may be limited to a few tens of kilohertz signal bandwidth, will nevertheless provide a much better set of LUT values than a simple statically based equivalent. The system in normal operation will be run using the direct DSP drive from the LUT; the speed of this will be mainly limited only by the speed of the DSP components. Unfortunately, as discussed in Chapter 3, such a system will still falter if the PA displays memory effects, although careful choice of a representative test signal may alleviate this limitation in some cases. Such "hybrid" combinations, consisting of a DSP-driven correction system which is operationally calibrated and updated by a closed analog loop would seem to have extensive scope.

Given such a relatively simple and convenient DSP calibration system, there is considerable justification for reconsidering an algorithmically based DSP correction system, depicted in Figure 5.27. Here the function of the DSP processor is to evaluate a suitable algorithm, based on the instantaneous envelope amplitude, in order to generate an appropriate correction drive to the vector modulator. This may appear to be an inherently slower process than the use of a ready-made look-up table; indeed, this may be a reason for LUT systems to dominate the current literature on this subject. But the advantage of an open loop system is that the signal itself can, in principle, be delayed for the duration of the DSP computation process; the software would have to be written such that the computation time would be approximately constant for all signal and correction levels. The key difference between an algorithmically based system and an LUT is that there may be only a few parameters required for the algorithm. It would, of course, be necessary to use more terms in a polynomial series than the simple third- and fifth-degree models which have been used throughout this chapter, to obtain the more demanding precision required in current MCPA specs. But even if

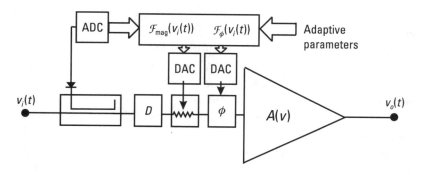

Figure 5.27 DSP predistortion scheme using algorithmic process to generate correction signals.

the number of parameters may be in the 10–20 range, this would be a much more manageable situation for adaptive control than a large LUT. Furthermore, the algorithm could incorporate some hysteresis, or memory, effects.

As DSP speed and availability increase, there seems little doubt that these techniques will play an increasing role in PA linearization.

5.7 Conclusions

Predistortion is a useful technique, which has possibly suffered from a lack of a satisfactory *a priori* design approach in the past. This chapter has shown one such approach, based on the inversion on the Volterra series model for the PA. The results from this analysis show some important practical guidelines and pitfalls for predistorter design. In particular, the traditional approach of tailoring a device with a "mirror image" gain expansion characteristic is both unsound and even unnecessary in many applications. The concept of a compound predistorter, where the various degrees of distortion can be accurately synthesized, with correct phasing angles, is a robust and little used approach to analog PD design. Its key feature is the elimination of the need to tailor the nonlinear characteristics of the PD device.

The availability of faster DSP has opened up possibilities for more precise realization of PD functions. It seems that DSP drivers will replace analog predistortion in most applications. There is some room for maneuver in the analog applicator, which is still required in a DSP PD system. This is conventionally viewed as a simple vector modulation device, but this raises issues of precision at well backed-off drive levels. A more satisfactory approach may be to take a compound analog PD device and use DSP to control the various amplitude and phase scaling adjustments, in a system which monitors the overall system performance in changing signal and environmental conditions.

On the negative side, PA memory and hysteresis effects represent a formidable limit to the ability of standalone PD systems to give the levels of correction possible using feedforward techniques. There is also a more fundamental problem that the cascading of two nonlinear devices leaves a residue of high-order nonlinear products that were absent in the original PA response. A predistorter has not only to linearize the target PA, but it has to clean up after itself as well. This aspect of predistortion remains an underrated problem, and has all too often been swept under the carpet by researchers who use carefully chosen spectrum sweep ranges to display their results. The simplified models used in this chapter can be criticized on the basis that they cannot accurately represent the nonlinear characteristics of some RFPAs. There is, however, a counterargument which says the analysis shows that such amplifiers cannot be successfully predistorted without excessive bandwidth in the video and digital drive circuits. The way out of this problem, and a path which appears to be followed by commercial PD-PA products, is to use a well-behaved PA which is well backed off from the compression region at PEP levels. This kind of amplifier lends itself not only to lower degree polynomial modeling, but more robust and useful predistortion linearization.

References

[1]   Schetzen, M., "Theory of pth Order Inverse of Nonlinear Systems," *IEEE Trans. on Circuits & Systems,* CAS-23, No. 5, 1976, pp. 285–291.

[2]   Yamakuchi, K., et al., "A Novel Series Diode Linearizer for Mobile Radio Power Amplifier," *Proc. IEEE Intl. Microw. Symp.,* MTT-S 1996, San Francisco, CA, WE3F-6, pp. 831–834.

[3]   Katz, A, et al., "Passive FET MMIC Linearizers for C, X, Ku-Band Satellite Applications," *Proc. IEEE Intl. Microw. Symp.,* MTT-S 1993, Atlanta, GA, WE3F-6, pp. 353–356.

[4]   Kumar, M., et al., "Predistortion Linearizer Using GaAs Dual-Gate MESFET for TWTA and SSPA Used in Satellite Transponders," *IEEE Trans. on Microwave Theory and Technology,* MTT-33, No. 12, 1985, pp. 1479–1499.

[5]   Nojima, T., et al., "The Design of a Predistortion Linearization Circuit for High Level Modulation Radio Systems," *Proc. Globecom,* 1985, pp. 47.4.1–47.4.6.

[6]   Ghaderi, M., et al., "Adaptive Predistortion Lineariser Using Polynomial Functions," *IEE Proc. on Communications,* Vol. 141, 1994, pp. 49–55.

[7]   Mino, J., and A. Valdovinos, "Amplifier Linearization Using a New Digital Predistorter for Digital Mobile Radio Systems," *Proc IEEE Conf. Vehic. Tech.,* VT-1997, pp. 671–675.

[8]   Cavers, J., "Amplifier Linearization Using a Digital Predistorter with Fast Adaption and Low Memory Requirements," *IEEE Trans. on Vehicular Technology,* VT-39, November 1990, pp. 374–382.

[9]   Cavers, J., "Optimum Indexing in Predistorting Amplifier Linearizers," *Proc. IEEE Conf. Vehic. Tech.,* VT-1997, pp. 676–680.

6

Feedforward Power Amplifiers

6.1 Introduction

Almost forgotten for a half century, the feedforward amplification technique has re-emerged as one of the most active technical topics in the wireless communication era. Despite continuing attempts to devise easier and more efficient alternatives, the feedforward method appears to be the most viable approach for making commercial PA products which can handle modern wideband multicarrier signal linearity specifications. Yet doubts remain about its production worthiness. It has to be classified as an open-loop correction method, and is therefore vulnerable to many of the effects for which closed-loop systems can claim some immunity. Environmental changes, drift in device and load characteristics, and even changes in the signal environment itself have to be carefully monitored by an extensive analog and digital housekeeping workforce, which can add substantially to the power consumption of a system which is already inefficient in its use of transistor periphery.

Such is the fundamental reliance of any practical feedforward system on monitoring and corrective adaption schemes, that the published literature of the last decade or so has almost entirely been concerned with these peripheral aspects. This includes a formidable litany of patents. But the real issue for final users, not to mention start-up entrepreneurs, seems to be the problem of converting the frequently admirable results obtained in research and engineering labs into an economically viable and producible product.

This chapter attempts to cover the theory of feedforward amplification in a manner that complements, rather than duplicates, existing in-depth treatments in recent literature [1, 2]. The approach, as with most of the main topics in this book, is to examine the operation of the system using simplified polynomial models for the PA components. This provides a quantitative analysis of feedforward operation and some of the tradeoffs between linearization performance and efficiency. The physical implementation of adaptive controls is not covered, but the theoretical treatment is able to make some useful statements about the potential of such controls. This is a back-to-basics approach, which hopefully yields a few mild surprises even to those already immersed in the design of commercial feedforward PA products.

6.2   The Feedforward Loop

Before looking at a block diagram (Figure 6.2), it is worthwhile first to consider Figure 6.1, which shows, in the simplest possible format, what a feedforward (FFW) loop actually does in relation to the PA within it. The basic action of the FFW loop is to provide, dynamically, the necessary power to "top-up" the gradually compressing characteristic of the main PA. In this respect it can be compared to a similar diagram (Figure 5.1) for a predistorter; the FFW loop is in effect an additive process which performs a similar linearization function, the difference being that the corrective action takes

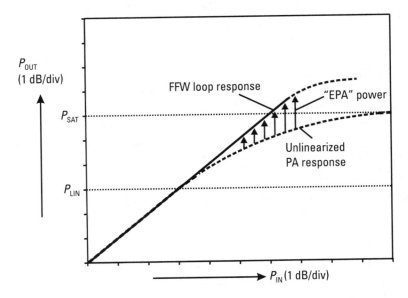

Figure 6.1 Basic action of feedforward linearization loop. For clarity, the loss of the output coupler has been ignored in the feedforward loop response.

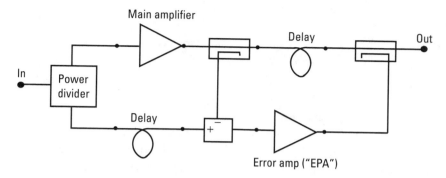

Figure 6.2 Basic feedforward loop.

place at the PA output. As with well-designed predistorters, the correcting signal will have a phase characteristic which also neutralizes any AM-PM distortion in the main PA. Clearly, this topping-up procedure has to be done by adding the required power at the PA output. One of the major frustrations of the FFW designer is the absence, and apparent impossibility, of a unidirectional, unequal power-combining device to perform this additive function. The process of power addition seems inevitably to involve the loss of main PA power and a need to generate the correction power at a much higher level than is originally required, based on the PA nonlinearity. This is an important issue and has a negative impact on the efficiency of feedforward systems. It will be discussed in more detail in Section 6.4.

A basic feedforward loop is shown in Figure 6.2. The operation of this loop is traditionally described by a vigorous display of hand-waving which all too frequently glosses over some important basic issues. We will therefore proceed in a quantitative manner at the outset, using the conventions established in previous chapters to simplify the analytical process. For a first pass, a simple cw RF carrier excitation will be assumed. As is customary in these analyses, AM-PM distortion will be ignored during the first pass, but will be reconsidered in due course.

Assuming, for further convenience, a symmetrical 3-dB input power divider, the input signal reaching the main PA is

$$v_{in} = v\cos\omega t$$

so the PA output, incorporating simple third-degree gain compression, is

$$v_{pa} = a_1(v\cos\omega t) - a_3(v\cos\omega t)^3 \tag{6.1}$$

then the signal appearing at the output of the sampling coupler on the PA output will be

$$\alpha v_{pa} = \alpha a_1 (v \cos \omega t) - \alpha a_3 (v \cos \omega t)^3$$

α being the voltage coupling factor.

This signal is now subtracted from the undistorted signal that has passed through the delay line, assumed lossless but having a signal delay matched to that of the PA. So the result of this subtraction gives an "error signal," v_e, given by

$$v_e = v \cos \omega t - \alpha v_{pa} = v \cos \omega t - \alpha a_1 (v \cos \omega t) + \alpha a_3 (v \cos \omega t)^3 \quad (6.2)$$

In practice, this subtraction process will inevitably include an extra attenuation factor on both signals; this is omitted here since it can be easily made up in the subsequent error amplifier gain.

This expression for v_e simplifies to something very useful-looking if $\alpha = 1/a_1$:

$$v_e = \alpha a_3 (v \cos \omega t)^3 \qquad (6.3)$$

Clearly, if this voltage is scaled back up by a factor of $1/\alpha (= a_1)$, and added to the signal emerging from the main PA given by (6.1), so that

$$v_{out} = a_1 (v \cos \omega t) - a_3 (v \cos \omega t)^3 + (1/\alpha)\alpha a_3 (v \cos \omega t)^3$$

the final output will be perfectly linearized. The error power amplifier (EPA) basically performs this scaling task; however, the gain of the error amplifier has to be greater than $1/\alpha$ in order to compensate for the (voltage) coupling factor β of the output coupler which is being used to achieve the necessary addition in the PA output. If, as has already been assumed for the present introductory treatment, the error insertion network is a conventional microwave directional coupler, the voltage coupling factor β will result in a complementary through-port transmission factor of $\sqrt{1 - \beta^2}$; this represents a tough design decision. For example, a value of $\beta = 0.5$, or 6-dB coupling factor, results in both a wastage of the main PA output power of 1.2 dB (= 10Log(4/3)), and a requirement to generate an additional 6 dB worth of correction signal. Lower values of the coupling factor β will result in higher

EPA power requirement, with less PA power wastage. Ever since the earliest reported implementations of feedforward systems [3], attempts have been made to reduce this power wastage.[1] This subject will be discussed further in Section 6.4.

This analysis assumes that the amplifiers and other components have sufficient bandwidth to include all of the nonlinear products which are generated by the main PA. In practice, the third harmonic components contained in the PA output (6.1) will be filtered out by the bandpass characteristics of the components. Fortunately, this does not affect the overall conclusion; the PA output will now contain only a first-order, third-degree gain compression term, obtained by expanding and band-limiting (6.1),

$$v_{pa} = a_1 v \cos \omega t - \frac{3}{4} a_3 v^3 \cos \omega t$$

which will still generate the appropriate first-order correction signal after passing through the differencing network, error power amplifier (EPA), and output insertion coupler.

Two critical issues will both complicate and degrade the simple linearization process as described so far. Firstly, the nonlinearity of the EPA needs to be considered. Secondly, the requirement for maintaining precise amplitude, delay, and phase tracking of the various signals around the loop needs also to be quantified. It is instructive to pursue the present, somewhat idealized, analysis to include both of these effects.

Considering the nonlinearity of the EPA first, it is clear that the EPA will always generate distortion products that will appear in the final loop output, and these will be completely outside the corrective action of the loop. This, unfortunately, is where the FFW loop shows a major distinction from closed loop feedback systems; the correction process is a one-way affair, which is not itself subject to further iterative revisions.[2] In order to pursue the effect of EPA nonlinearity on the loop linearization performance, it is necessary to make a design decision concerning the power capability of the

---

1. It is worth further note here that if the main PA is at its 1-dB compression point, and the FFW loop restores linear operation, a 1-dB through loss in the coupler has not reduced the power available from the PA output transistors from that which can be obtained in an unlinearized configuration; this point was strongly made by Seidel [3].

2. This is a statement in support of the double-loop FFW system, which is considered later in this chapter.

EPA, in relation to the main PA. Clearly, it is desirable to have a much lower power EPA in order to maximize the system efficiency, but the output level at which the system linearity becomes limited by EPA distortion will then also be lower. Selection of EPA power develops quickly into a complicated design problem, interacting with the power backoff (PBO) level of the main PA, the choice of both coupling factors (α and β), and the fidelity of amplitude and delay tracking. This tradeoff will be considered further in Sections 6.5 and 6.6. For the present analysis, the EPA is assumed to have a simple third-degree characteristic defined by power series coefficients [b_n], from which some preliminary observations can be made concerning the effects of the EPA on loop performance.

Picking up the analysis from the expression for v_e (6.3), the EPA output will be

$$v_{epa} = b_1 v_e - b_3 v_e{}^3$$
$$= b_1 \left(\alpha a_3 v^3 \cos^3 \omega t \right) - b_3 \left(\alpha a_3 v^3 \cos^3 \omega t \right)^3 \tag{6.4}$$

As already observed, the first term in (6.4) will be the originally desired cancellation voltage which linearizes the main PA output, assuming that the EPA gain b_1 incorporates both the sampling factor ($1/\alpha$) and the output insertion factor ($1/\beta$). The second term clearly has a large number of higher harmonic components, all of which can be assumed to be outside the system bandwidth. There is, however, a residual fundamental component which is proportional to the third power of the main PA third-degree coefficient a_3 and, significantly, also proportional to the ninth power of the input signal magnitude, v^9. This term represents uncorrected output gain compression caused by the nonlinearity of the EPA. The 9:1 PBO of residual distortion products that will be displayed by such an FFW loop is an important factor in determining the required EPA power capability, since there is little point in reducing the EPA nonlinear contribution to a level which is lower than the signal resulting from a cancellation error between the third-degree terms in the EPA and PA outputs. The dependency of the residual distortion term on the third power of a_3 is also highly significant. This shows that any reduction in a_3, corresponding for example to a PA provided with a predistorter, an internal feedback loop, or even an internal feedforward loop, will show a greatly magnified improvement when placed into an FFW system.

The above analysis serves as an introduction to the operation of a feedforward loop. For any selected signal amplitude v, the PA distortion products are isolated by the process of sampling and subtraction in the first loop, followed by cancellation in the second loop. Clearly, this is a "fast" process, performed entirely in the RF time domain, without any need to convert the signals to baseband or IF. Thus if the amplitude v in the above analysis is now made a time-varying envelope function $v(\tau)$, the loop can be expected to respond and maintain an output envelope which is a highly linear replica of the input envelope function. Indeed, the timescale on which τ may vary, and the loop integrity maintained, will be much faster than most of the alternative linearization methods so far discussed. In this sense, the above analysis can be considered to be a more general case as far as the excitation is concerned. It would be appropriate, however, to include the possibility of AM-PM distortion in the main PA.

If the third-degree Volterra phase angle is now included in the PA output, the output from the main PA can be rewritten as

$$v_{pa} = a_1 \{ v \cos \omega t \} + a_3 \{ v \cos (\omega t + \varphi_3) \}^3 \qquad (6.5)$$

which will thus show AM-PM distortion. As discussed in Chapter 3, the value of a_3 will increase, for a stipulated level of gain compression, as the angle φ_3 takes on values other than the zero AM-PM case of $\varphi_3 = \pi$. Thus the required EPA power output for linearizing the amplifier will increase due to the presence of AM-PM, at a given level of AM-AM, or gain compression. This effect can be quantified by expanding (6.5),

$$v_{pa} = a_1 \{ v \cos \omega t \} + \left(\frac{3}{4} \right) a_3 v^3 \{ \cos (\omega t + \varphi_3) \} \qquad (6.6)$$

where only first-order, or fundamental, terms have been retained.

Thus, the error signal v_e at the output of the differencing network will be

$$v_e = -\alpha \left(\frac{3}{4} \right) a_3 v^3 \{ \cos (\omega t + \varphi_3) \}$$

It is already clear that this is the required correction signal for addition to the PA output, after suitable scaling. But the increase of a_3, which follows directly from non-π values of the Volterra angle φ_3, means that the EPA has to generate more power for a main PA which has AM-PM at a given level of gain compression and at a given output power. This is an important issue and is worthy of closer inspection.

6.3 AM-PM Correction in the Feedforward Loop

It will be more convenient in this section to describe the PA output in terms of the measured AM-PM at a given input amplitude level, since this is the usual manner in which AM-PM distortion is measured and specified. So returning once again to the expression for the PA output, (6.1), this will now take the form

$$v_{pa} = \gamma v \cos(\omega t + \phi) \qquad (6.7)$$

where γ represents the fractional voltage gain compression and ϕ the AM-PM distortion. This formulation seems quite natural from a pragmatic standpoint, since it uses directly measurable PA characteristics. It is, however, hazardous from an analytical viewpoint in that γ and ϕ are both functions of the input signal amplitude v. It nevertheless serves the present purpose, which is to quantify the additional EPA power requirements in the presence of AM-PM.

After passing through the error-signal forming loop, the resulting correction signal generated at the output will have the form

$$v_{cor} \cos(\omega t + \Delta) = v \cos \omega t - \gamma v \cos(\omega t + \phi) \qquad (6.8)$$

where v_{cor} represents the amplitude of the EPA output signal after it has passed through the output insertion coupler. Clearly, if $\gamma = 1$ and $\phi = 0$, there is no correction signal.

Equation (6.8) rearranges to give

$$v_{cor} \cos \Delta = v(1 - \gamma \cos \phi)$$
$$v_{cor} \sin \Delta = \gamma v \sin \phi$$

so that

$$v_{cor}^2 = v^2 \left(1 + \gamma^2 - 2\gamma \cos \phi \right) \qquad (6.9)$$

The formulation of (6.9) contains some surprises. A frequently asked question about PA distortion is the relative deleterious effects of AM-PM in comparison to gain compression. Equation (6.9) allows a direct comparison to be made, using the EPA power as a quantitative measure of discomfort, in the context of a feedforward correction scheme. For example, a PA which has no

AM-PM ($\phi = 0$) will require an EPA correction signal at its 1-dB compression point, given by

$$v_{c1dB}^{2} = v^{2}(1-\gamma)^{2}$$
$$v_{c1dB} = v\left(1-\left(10^{-0.05}\right)\right) \qquad (6.10)$$
$$= (0.1087)v$$

Alternatively, a PA which has AM-PM but no gain compression ($\gamma = 1$), at the same input level v, requires an EPA correction voltage given by

$$v_{c\phi}^{2} = 2v^{2}(1-\cos\phi) \qquad (6.11)$$

For example, if the AM-PM is 10°, (6.11) has a value of

$$v_{c10}^{2} = 2v^{2}\left(1-\cos(10^{0})\right)$$
$$v_{c10} = (0.1743)v \qquad (6.12)$$

which can be seen to be a bigger signal than that required for the 1-dB compression case, actually about 4 dB higher in power. In fact, the AM-PM which gives the same level of EPA correction as that in (6.10) is about 6°.

In practice, of course, PAs have both gain compression and AM-PM. The effect of the AM-PM is to place a significant extra workload on the EPA, as shown in the vector diagram of Figure 6.3. Clearly, as the PA output signal shifts in phase around a constant radius representing a particular compression level, the length of the correction vector increases rapidly, even for lower compression values. The "error vectors" shown in Figure 6.3 demonstrate a weakness, possibly even an Achilles' heel, in the otherwise admirable feedforward correction process. It is one thing to provide an external power source in order to top-up the gain compression of a PA; this is entirely within normal intuitive expectations based on energy considerations. But the use of additive power to correct a phase error, albeit in a mathematically precise manner, seems wasteful. In designing and selecting a PA for feedforward use, it is clearly important to check the AM-PM response; anything over about 5° in the intended operating power range will start to dominate the gain compression in terms of the EPA power requirement. It may well be argued that the intention is to operate the PA well backed-off from the 1-dB compression point. But even in the backed-off region, corresponding to the lower

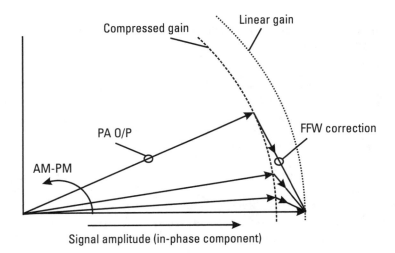

Figure 6.3 Effect of AM-PM on feedforward loop correction signal.

compression levels not shown in Figure 6.3, there is still a very real possibility of just a few degrees of AM-PM seriously upsetting the EPA power budget calculations. This is worth one further step in the analysis, which is to return to the PA Volterra series formulation to obtain a realistic coupling between the gain compression and AM-PM values in a power backoff situation.

Returning then to (6.6), the PA output can be expressed as

$$v_{pa} = a_1 \{v \cos \omega t\} + \left(\frac{3}{4}\right) a_3 v^3 \{\cos(\omega t + \varphi_3)\}$$

so the relationships between the PA parameters a_1, a_3, φ_3 and the fractional voltage gain compression factor γ and AM-PM ϕ, at a given input amplitude v, have been shown (see Chapter 3) to be

$$\gamma a_1 v \cos \phi = a_1 v - \left(\frac{3}{4}\right) a_3 v^3 \cos \varphi_3$$

$$\gamma a_1 v \sin \phi = \left(\frac{3}{4}\right) a_3 v^3 \sin \varphi_3 \qquad (6.13)$$

We may assume for the purposes of this calculation, that a_3 is normalized to a value of $a_1 = 1$, and that the 1-dB compression point occurs at $v = 1$. It is then possible to derive pairs of values of a_3 and φ_3 for selected levels of AM-PM ($\phi_{1\,dB}$) at the 1-dB compression point,

$$a_3 = \tfrac{4}{3}\sqrt{1+\gamma^2 - 2\gamma\cos\phi_{1\,\mathrm{dB}}}$$

$$\tan\varphi_3 = \frac{\gamma\sin\phi_{1\,\mathrm{dB}}}{1-\gamma\cos\phi_{1\,\mathrm{dB}}}$$

For example, for a PA having 10° AM-PM at its 1-dB compression point, $a_3 = 0.263$, $\varphi_3 = 128°$. The EPA voltage correction, normalized, as before, for $a_1 = 1$ is given (from 6.6) by

$$v_e = -\left(\frac{3}{4}\right)a_3 v^3 \left\{\cos\left(\omega t + \varphi_3\right)\right\}$$

so that the EPA correction can be redrawn as a function of power backoff, using chosen values for a_3, and φ_3. Clearly, the function shows a 3:1 power backoff with drive signal amplitude, but the key issue is that each backoff curve shows an upward offset as φ_3 rotates away from the 180° value which corresponds to a PA having no AM-PM (Figure 6.4). The key point about Figure 6.4 is that it shows the deleterious effect of AM-PM on the EPA power requirement persists with power backoff; it is not restricted to the upper end of the signal range where AM-PM becomes directly measurable.

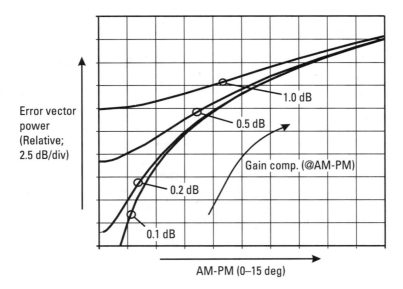

Figure 6.4 Effect of AM-PM on error vector magnitude at low levels of gain compression.

The analysis in this section appears to have identified an important, and sometimes overlooked, issue in the design of RFPAs for feedforward applications. Too often, the AM-PM specification is regarded as being of secondary, rather than primary, importance. This is mainly a problem of convention, that the 1-dB compression point of an amplifier essentially specifies its power and linearity performance. The concept of error vector magnitude (EVM) defined and discussed in Chapter 3, would appear to be a much better and more generalized way of specifying PA linearity.

6.4 Error Insertion Coupling

It is clear from the above introductory analysis that the error insertion coupler is a key element in a feedforward loop. It is the last element in the power chain, and performs its task in an open-loop fashion. It also appears to waste much valuable energy, both from the EPA and the main PA itself. For this reason, the choice of the error insertion coupling coefficient β has become something of a *cause célèbre* amongst FFW researchers and writers. The practical range would appear intuitively to be about 6–10 dB; there appears to be a tradeoff between the higher transmission loss of lower β values and the higher EPA requirement predicated by higher values. The apparent need to waste energy in this manner has caused much comment and focused research, and the subject is worthy of more detailed scrutiny prior to a more generalized analysis of a feedforward loop.

A microwave coupler is a familiar item but its properties can sometimes be misrepresented. As a multiport passive device, linear voltage and current superposition have to apply, but must also demonstrably comply with energy conservation. In the laboratory a coupler is widely perceived as a passive device which superimposes powers, rather than voltages, at its ports. All of these observations can be reconciled, so long as due attention is paid to the amplitudes and phasing of the signals at the various ports.

Figure 6.5 shows a directional coupler with all four ports terminated and a single sinusoidal signal applied to port 1. Assuming that this is a well-designed high directivity coupler, it can be assumed that the even- and odd-mode impedances follow the classical relationship

$$Z_{ev} Z_{odd} = Z_o^{2}$$

where Z_o is the termination impedance, and that the even- and odd-mode propagation velocities are equal. At the frequency corresponding to an electrical quarter wavelength, the voltages at the passive ports are given by

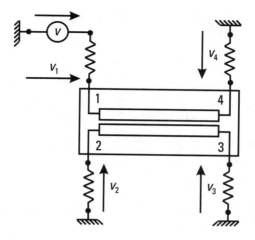

Figure 6.5 Terminated directional coupler with single sinusoidal signal excitation.

$$V_2 = \beta v$$
$$V_3 = 0$$
$$V_4 = -j\gamma v$$

where the coupling coefficient β is given by

$$\beta = \left(\frac{Z_{ev} - Z_{odd}}{Z_{ev} + Z_{odd}} \right)$$

and the transmission coefficient γ is given by

$$\gamma^2 = 1 - \beta^2$$

which clearly ensures that energy is conserved for a single signal excitation.

Figure 6.6 shows the same coupler, with cophased sinusoidal signals v_1 and v_3 applied to ports 1 and 3, respectively. By the unimpeachable law of linear superposition, the voltages at the passive ports will now be

$$V_2 = \beta v_1 - j\gamma v_3$$
$$V_4 = -j\gamma v_1 + \beta v_3$$

(6.14)

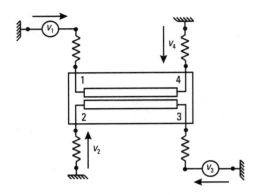

Figure 6.6 Terminated directional coupler with two sinusoidal cophased input signals.

Here lies the essence of the "coupler misconception." Due to the quadrature voltage relationships, there is no violation of power conservation; assuming a unity impedance environment, the output power can be expressed as

$$P_2 = \beta^2 v_1^2 + \gamma^2 v_3^2$$
$$P_4 = \gamma^2 v_1^2 + \beta^2 v_3^2$$
$$P_2 + P_4 = v_1^2 + v_3^2$$

which is the input power to ports 1 and 3.

In order to use such a coupler as a means of adding the two signals at port 4 in a phase-coherent manner, it is necessary to place a delay on the signal input to port 3, which equals the direct path delay from port 1 to port 4, as shown in Figure 6.7. In the specific quarter-wave case, this can be represented by a $-j$ multiplier on v_3 in (6.14) giving

$$V_2 = \beta v_1 - \gamma v_3$$
$$V_4 = -j(\gamma v_1 + \beta v_3)$$

(6.15)

so that

$$P_2 = \beta^2 v_1^2 + \gamma^2 v_3^2 - \beta\gamma 2 v_1 v_3$$
$$P_4 = \gamma^2 v_1^2 + \beta^2 v_3^2 + \beta\gamma 2 v_1 v_3$$
$$P_2 + P_4 = v_1^2 + v_3^2$$

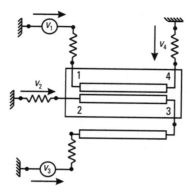

Figure 6.7 Directional coupler used as a signal combiner.

and once again energy is conserved. In practice, however, the partial signal cancellation at port 2 is rarely measured; the "dump" port termination is often concealed within the coupler body.[3] It is however this cancellation which explains some further misconceptions which can arise in using couplers as power combiners, as distinct from voltage adders. It should be noted that at frequencies displaced from the quarter-wave setting, the in-phase addition at port 4 will be maintained so long as the delay section tracks the coupler length. The voltage components at port 2 will not now appear in perfect antiphase, but the amplitude offsets in the coupler response will ensure that energy is still conserved.

An application for the coupler configuration of Figure 6.7, which is highly relevant to FFW systems applications, is shown in Figure 6.8. The shortfall of output power from a PA is being restored by the use of a lower power "auxiliary" PA. Clearly, for this scheme to be compared directly with an uncompensated PA, the auxiliary PA should supply enough power to compensate the PA compression and *also* the direct transmission factor of the coupler. The objective, therefore, is to select a coupling ratio β which enables the task to be performed using the lowest auxiliary PA. Taking the amplifier output compression as ε dB, the PA output can be written as

$$v_{pa} = 10^{-\varepsilon/20}\, v$$

where v is the desired uncompressed output level.

---

3. Some coupler manufacturers provide an external coaxial termination for this "unused" port, but the user is mysteriously discouraged from using it through the liberal application of paint on the connector pair.

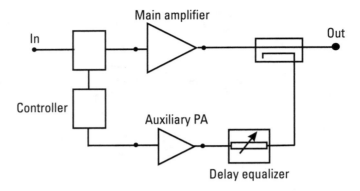

Figure 6.8 Use of auxiliary PA to restore gain compression in main PA.

So the voltage level required from the auxiliary PA can be expressed as

$$v = \gamma v_{pa} + \beta v_{aux}$$
$$= \gamma 10^{-\varepsilon/20} v + \beta v_{aux}$$

giving

$$\frac{v_{aux}}{v} = \frac{\left(1 - \gamma 10^{-\varepsilon/20}\right)}{\beta} \tag{6.16}$$

with $\gamma = \sqrt{1 - \beta^2}$.

The relationship of (6.16) can be plotted out as a decibel ratio for the powers of the two amplifiers against the coupling factor, also expressed conventionally in decibels; this is shown in Figure 6.9. The plot for 1-dB compression may come as something of a surprise. At the optimum coupling ratio of 7 dB, the auxiliary PA power required to restore the compression of the main PA and also compensate for the (approximately) 1-dB transmission factor of the coupler is 6.87 dB lower than the linearized PA output, or 5.87 dB lower than its actual 1-dB compressed power output.

This is a much more modest auxiliary PA than is sometimes speculated, using erroneous reckoning based on power rather than voltage compensation. Such an argument might run:

PA output at 1-dB compression point, 100W;

Linearized PA output with compression removed, 125.9W;

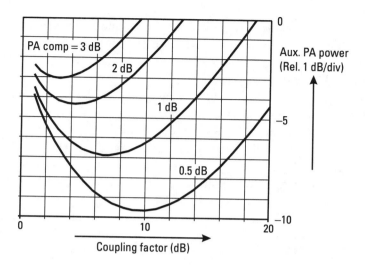

Figure 6.9 Relative power required to restore PA compression using the system in Figure 6.8.

Output from 7-dB coupler with no auxiliary PA, 80W;

Compensation power required from auxiliary PA, 46W;

Power required from auxiliary PA (7-dB coupling), $46 \times 5 = 230W(!)$.

The correct reasoning, based on voltage addition at the coupler output port, gives a power which is 6.87 dB lower than 100W, or **25.9W**. Obviously, to generate this power the auxiliary PA would have to have a 1-dB compression power somewhat higher than this value. What makes this result surprising is the much more widespread belief that power is wasted in the coupler dump port. Using the same numerical example, surely we have to accept that a 7-dB chunk of the PA output, 20W, is immediately dumped into port 2? Not only that, but a large proportion of the auxiliary PA power, 20.7W, is also transmitted wastefully to port 2. Once again, this is a misconception which can be relieved by considering (6.14). The phase relationships of the coupler put these two components in antiphase at port 2, so the voltage levels subtract. In this particular case, taking voltages to be the square roots of the power levels (i.e., unity impedance level), the voltage at port 2 is

$$V_2 = \sqrt{20.7} - \sqrt{20}$$
$$P_2 \approx 0.006W$$

which allowing for some arithmetic imprecision (and the optimum coupling value in Figure 6.9 not being precisely 7.0 dB) is compatible with the energy of the auxiliary PA (25.9W) entirely adding to the energy of the PA (100W) to give a final output power of 125.9W.

In the above calculation, the zeroing of P_2 actually corresponds to the choice of the minimum point on the coupling curve shown in Figure 6.9. If the optimum coupling value is used, the coupler will restore the stipulated input power shortfall (1 dB in this case) with no wastage of energy from either source. In this mode, the coupler is behaving as an asymmetrical, lossless power combiner; none of the above results would be found surprising had the input shortfall of power been taken as 3 dB and a 3-dB coupling factor found to be the lossless optimum value. Such a configuration can be easily recognized as a conventional quadrature power combiner. The asymmetrical version can be designed using (6.15), and setting the port 2 output to zero, giving

$$v_2 = \beta v_1 - \gamma v_3 = 0$$

$$\frac{v_3}{v_1} = \frac{\beta}{\gamma} \tag{6.17}$$

$$\frac{P_3}{P_1} = \frac{\beta^2}{\gamma^2} = \frac{1-\gamma^2}{\gamma^2}$$

thus any two unequal cophased signals can be losslessly combined using a coupling ratio as prescribed in (6.17).

There is a further consideration on these results when they are applied to an FFW loop. It has so far been assumed that the auxiliary PA must overcome the coupler transmission factor as well as neutralizing the gain compression of the main PA. An FFW loop in normal adjustment monitors the PA output before, rather than after, the error insertion coupling. The loop therefore generates a correction signal which does not account for the coupler transmission factor γ. There is in addition a convention, or tradition, amongst FFW system designers to accept the coupler transmission as just one of several output chain loss contributors. This approach is reinforced later when the possibility of using the EPA to restore the coupler transmission factor is further considered. The shortfall of power basically has to be made up by using larger power transistors. With this revised criterion, the curve of Figure 6.9 can be recalculated, using the modified form of (6.16):

$$\frac{v_{aux}}{v} = \frac{\gamma}{\beta}\left(1 - 10^{-\varepsilon/20}\right) \tag{6.18}$$

shown plotted in Figure 6.10. Clearly, if the coupler transmission is swept under the carpet in this fashion, there is a preference for lower coupling factors in the sense that lower auxiliary power is required. Values below 10 dB, however, are judged to be excessively wasteful on PA transistor periphery and this value is frequently taken to be a satisfactory optimum. Returning to the specific numbers in the chosen example, and a 10-dB coupling factor, the auxiliary PA, or EPA, has to generate a power level of

$$\frac{0.9}{1/10}\left(1 - 10^{-1/20}\right)^2 = 0.106$$

or −9.7 dB relative to the linearized PA output in order to restore the output power at the 1-dB compression point. Although this is 3 dB lower than the power required in the comparable calculation above for complete power restoration using an optimum 7-dB coupling factor, later analysis will show that this configuration still justifies preference in a feedforward system. So the 100W output is attenuated by the coupler transmission of 0.46 dB, down to 90W, and then restored to 1-dB higher power of 113W by the coupled EPA power. The EPA power is $0.106 \times 126 = 13.3$W. Once again, the "wastage" of power is minimal: $(100 + 13.3) - 113 = 0.3$W. Unfortunately, at lower power levels the auxiliary PA is not required to provide anything but the

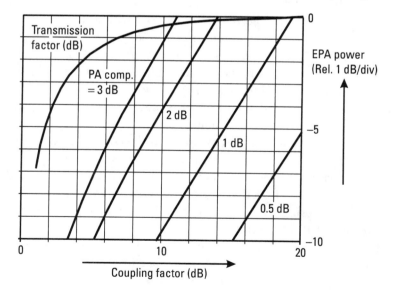

Figure 6.10 Auxiliary PA requirement, relative to uncompressed main PA output; coupler transmission "accepted" as uncompensated loss (see text).

tiniest amount of power to perform a linearization function, and the coupler transmission factor γ reverts to just another loss in the output power chain. The analysis in the first part of this section suggests that a more optimum arrangement may be to adjust the feedforward loop such that the EPA generates a signal which restores the coupler transmission loss as well as linearizing the PA output. Section 6.5.3 will consider this option quantitatively.

6.5 Third-Degree Analysis of the Generalized Feedforward (FFW) Loop

6.5.1 Formulation and Analysis

Throughout this book, polynomial models with various levels of simplification have been used to provide quantitative information about the nonlinear behavior of an RF amplifier. Such models, especially in truncated forms, can only replicate the behavior of practical RF amplifiers approximately. Chapter 3 examines this issue in considerable depth, showing that several higher degrees of linearity, perhaps extending up to the ninth or eleventh, are required to model typical Class AB amplifiers up to, and beyond, the 1-dB compression point. Accepting that there is a clear distinction between an accurate model for CAD simulation purposes, and a model which is approximate but enables useful overall performance characteristics to be analyzed in a quantitative fashion, it is justifiable to proceed with a full analysis of the feedforward loop using a simple third-degree model for both RFPAs.

This analysis will enable some quantification on various tradeoffs between EPA power, power backoff, tracking fidelity, and overall efficiency. The analysis will also reveal that there is an additional parameter of great significance in the design and alignment of practical feedforward systems; the first loop coupling factor α. This parameter may not necessarily be set to cancel the linear gain of the main PA exactly, and the chosen value of α has an important impact on the loop performance and overall efficiency. To keep the analysis within manageable limits, AM-PM effects have not been included. The results of Section 6.4 can be used in conjunction with the results of the foregoing analysis in order to determine the increases in EPA power which AM-PM demands.

The system analyzed is shown in Figure 6.11. The main PA has power series coefficients a_1, a_3, and the EPA b_1, b_3. The first loop coupling factor is α, and the output error insertion coupling factor is β. The coupling factor β forces an output transmission factor of γ, where $\gamma = \sqrt{1 - \beta^2}$. No resistive

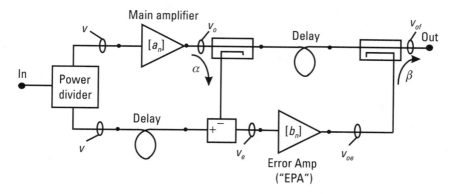

Figure 6.11 Feedforward loop analysis.

losses in the components are included in the analysis. A parameter of key significance in this analysis will be the power capacity of the EPA. For convenience, this will be expressed as a simple ratio to the main PA. This ratio will be termed the error PA ratio, or *EPR*, and is the ratio of the main PA 1-dB compression power to that of the EPA. It follows from the power series that the EPR (E) can be expressed in terms of the a_n and b_n coefficients:

$$E = \left(\frac{a_1}{b_1}\right)^3 \left(\frac{b_3}{a_3}\right) \qquad (6.19)$$

This expression will be used primarily to determine the value of b_3 for a prescribed EPR level:

$$b_3 = a_3 \left(\frac{b_1}{a_1}\right)^3 E \qquad (6.20)$$

the b_1 gain value will be determined primarily from the product of the coupling factors α and β, but will also be used as a convenient gain tracking adjustment. In general, it should be noted that the Greek symbols are used to represent voltage attenuation factors less than unity. The analysis will be performed using a sinusoidal RF carrier of prescribed amplitude v. Once suitable expressions for the various inputs and outputs around the loop have been derived as polynomial functions of the "quasi-static" carrier amplitude v, a suitable modulation function $v(\tau)$ can be considered.

The output from the main PA is

$$v_o = a_1 v \cos \omega t + a_3 v^3 \cos^3 \omega t$$

$$= \left\{ a_1 v + \frac{3}{4} a_3 v^3 \right\} \cos \omega t + \frac{1}{4} a_3 v^3 \cos 3\omega t$$

which making the usual band-limited assumption reduces to

$$v_o = \left\{ a_1 v + a_3 \frac{3}{4} v^3 \right\} \cos \omega t$$

The input to the error amplifier (EPA) is then

$$v_e = \left\{ v - \alpha \left(a_1 v + \frac{3}{4} a_3 v^3 \right) \right\} \cos \omega t \qquad (6.21)$$

In what will be termed "normal adjustment" of the FFW loop, the coupling factor α will be set to cancel the linear gain of the main PA:

$$\alpha = \frac{1}{a_1}$$

so that the EPA receives a signal

$$v_e = \left\{ -\frac{3}{4} \alpha a_3 v^3 \right\} \cos \omega t$$

This signal, as already observed in the discussion following (6.4) earlier, results in a ninth-order dependency distortion product at the FFW output which is not corrected by the system. An alternative adjustment of the system, called "compression adjustment" can be used to reduce the input to the EPA at higher levels of drive amplitude v. Basically, (6.21) can be solved, for any selected value of v, to give a value for α which zeroes the EPA input,

$$\alpha = \frac{v}{a_1 v + \frac{3}{4} a_3 v^3} \qquad (6.22)$$

For example, at the 1-dB compression point, the value of α would correspond to an attenuation level which cancelled the main PA gain minus 1dB. In this mode, the FFW loop will provide an error signal which will always set the overall gain to the compressed level. At the selected compression level, the EPA receives no input and will not generate any significant distortion in the FFW output. This opens up possibilities for reducing the power of the EPR, but has some additional side effects which may be considered to be less desirable. This is a subject of much debate and will be a central issue in the present analysis. Continuing from the expression for EPA input in (6.21), the EPA output can be expressed as

$$v_{oe} = b_1 \left\{ v - \alpha \left(a_1 v + \frac{3}{4} a_3 v^3 \right) \right\} \cos \omega t + b_3 \left\{ v - \alpha \left(a_1 v + \frac{3}{4} a_3 v^3 \right) \right\}^3 \cos^3 \omega t$$

which can be band-limited to

$$v_{oe} = \left[b_1 \left\{ v - \alpha \left(a_1 v + \frac{3}{4} a_3 v^3 \right) \right\} + \frac{3}{4} b_3 \left\{ v - \alpha \left(a_1 v + \frac{3}{4} a_3 v^3 \right) \right\}^3 \right] \cos \omega t$$

After some manipulation the $\cos \omega t$ coefficient can be rewritten in the form

$$|v_{oe}| = \{ b_1 (1 - \alpha a_1) \} v + \left\{ -\frac{3 b_1 \alpha a_3}{4} + \frac{3 b_3}{4} (1 - \alpha^3 a_1^3) + 3 \alpha a_1 (\alpha a_1 - 1) \right\} v^3$$

$$+ \left\{ \frac{27 b_3 \alpha a_3}{16} (2 \alpha a_1 - \alpha^2 a_1^2 - 1) \right\} v^5 + \left\{ \frac{81 b_3 a_3^3 \alpha^2}{64} (1 - \alpha a_1) \right\} v^7$$

$$+ \left\{ -\frac{81 b_3 \alpha^3 a_3^3}{256} \right\} v^9 \tag{6.23}$$

In the "normal adjustment" case, $\alpha a_1 = 1$, leaving

$$v_{oe} = \left[-\left\{ \frac{3 b_1 \alpha a_3}{4} \right\} v^3 - \left\{ \frac{81 b_3 \alpha^3 a_3^3}{256} \right\} v^9 \right] \cos \omega t$$

Thus, the output from the whole loop,

$$v_{of} = \gamma v_o + \beta v_{oe}$$

where β is the output insertion coupling factor, and γ is the corresponding transmission factor through the coupler $\left(\gamma = \sqrt{1 - \beta^2}\right)$, is

$$v_{of} = \left[\gamma\left(a_1 v + \frac{3a_3}{4}v^3\right) + \beta\left(-\frac{3b_1\alpha a_3}{4}v^3 - \frac{81b_3\alpha^3 a_3^3}{256}v^9\right)\right]\cos\omega t$$

(6.24)

so putting

$$\beta b_1 \alpha = \gamma$$

(6.25)

the original third-degree PA distortion is cancelled, and the output from the normally adjusted FFW loop is

$$v_{of} = \gamma a_1 v - \beta\frac{81}{256}b_3\alpha^3 a_3^3 v^9$$

(6.26)

In order to evaluate (6.26) quantitatively, it is necessary to set values for β, b_1, and b_3.

For the error coupling we will opt for the lower transmission loss option of 10 dB $\left(\beta = 1/\sqrt{10}\right)$, as discussed in Section 6.4. Given an established value for β, (6.25) gives a value for the EPA gain b_1, and b_3 can be determined using (6.20), for a specified EPR level E. The original PA power series coefficients a_1 and a_3 can now be normalized conveniently; the gain a_1 is normalized to unity and the a_3 value is chosen such that the 1-dB compression point is reached at an input sinusoidal signal amplitude $v = 1$:

$$10^{-0.05}v = v - \frac{3}{4}a_3 v^3$$

with $v = 1$, $a_3 = 0.145$.

This calculation for a_3 obviously assumes no AM-PM. Section 6.3 showed that the main effect of AM-PM is to increase the value of a_3 for a given level of compression. In this sense, the effect of AM-PM can still be estimated in this analysis, simply by modifying the value of a_3. The simulation in Section 6.6 will further quantify the effects of AM-PM in a FFW loop.

We can now define a suitable modulation function $v(\tau)$. For the present purposes, a two-carrier signal will be selected, so that

$$v(\tau) = V \cos \Omega t$$

where Ω is a "modulation domain" frequency, and will be at least two orders of magnitude smaller than ω. The above analysis presents the various output voltages in the form

$$v_o = \left(A_1 v + A_3 v^3 + A_5 v^5 + A_7 v^7 + A_9 v^9 \right) \cos \omega t \qquad (6.27)$$

where the A_n coefficients are functions of the loop parameters α, β, γ, and the two PA coefficients a_n and b_n. With the defined time variation of v,

$$v(\tau) = V \cos \Omega t$$

(6.27) becomes

$$v_o = \begin{pmatrix} A_1 V \cos \Omega t + A_3 V^3 \cos^3 \Omega t + A_5 V^5 \cos^5 \Omega t \\ + A^7 V^7 \cos^7 \Omega t + A_9 V^9 \cos^9 \Omega t \end{pmatrix} \cos \omega t \quad (6.28)$$

If the cosine polynomial powers are expanded into harmonic series (see Chapter 3), it is possible to rearrange (6.28) into the form

$$v_o = \begin{cases} \left(A_1 V + \dfrac{3}{4} A_3 V^3 + \dfrac{5}{8} A_5 V^5 + \dfrac{35}{64} A_7 V^7 + \dfrac{63}{128} A_9 V^9 \right) \cos \Omega t \\[2mm] + \left(\dfrac{1}{4} A_3 V^3 + \dfrac{5}{16} A_5 V^5 + \dfrac{21}{64} A_7 V^7 + \dfrac{21}{64} A_9 V^9 \right) \cos 3\Omega t \\[2mm] + \left(\dfrac{1}{16} A_5 V^5 + \dfrac{7}{64} A_7 V^7 + \dfrac{9}{64} A_9 V^9 \right) \cos 5\Omega t \\[2mm] + \left(\dfrac{1}{64} A_7 V^7 + \dfrac{9}{256} A_9 V^9 \right) \cos 7\Omega t \\[2mm] + \left(\dfrac{1}{256} A_9 V^9 \right) \cos 9\Omega t \end{cases} \cos \omega t$$

$$(6.29)$$

so that the inner bracketed expressions represent the relative magnitude of the various orders of IM products (an additional factor of $\frac{1}{2}$ in each case is required to generate the magnitude of an individual upper or lower IM sideband).

Figure 6.12 shows a plot of the third-order IM distortion for the prescribed two-carrier excitation of the FFW system in normal adjustment ($\alpha = 1$), compared with the uncorrected PA. A ninth-degree distortion product backoff characteristic is displayed, showing the effect of EPR selection. These characteristics assume that the system, and especially the value of b_1 defined in (6.25), is precisely maintained over all time and conditions. Clearly, in practice there will always be a level of precision to which the required tracking can be maintained.

6.5.2 Tracking Errors

Equation (6.24) presents a more generalized expression for the FFW loop output which allows the effects of offset tracking to be examined. Figure 6.13 shows the result of a range of "second-loop" tracking errors on the IM3 output, using the median EPR value of 10 dB from Figure 6.12. Such tracking errors simulate the effects of imperfect error insertion at the output coupler. Figure 6.13 shows a ±0.25-dB range of tracking errors. The effect of such

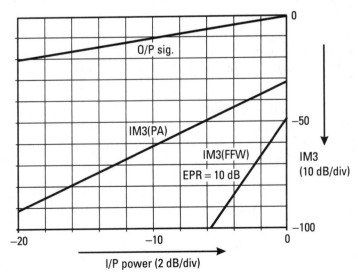

Figure 6.12 Two-carrier IM3 response for perfectly tracked FFW loop; EPR = 10 dB. 0-dB input power corresponds to 1-dB gain compression at the peak envelope power (PEP).

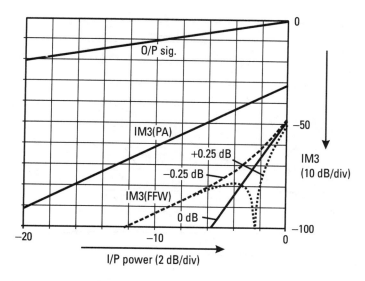

Figure 6.13 Effect of ±0.25 dB tracking error on FFW loop; parameters as defined in Figure 6.12.

tracking errors is to reduce the backoff slope from 9:1 down to 3:1, the breakpoint being approximately the point where the residual third-degree error signal equals the ninth-degree EPA distortion signal. Clearly, this breakpoint is a function of the EPR value as well as the tracking error itself.

Figure 6.13 shows a surprising effect when the error signal is slightly higher than the ideal value. The FFW loop distortion output now goes through a null which can be moved around in the PBO range by suitable choice of tracking error. At the higher end of the power range, this null can actually reduce the FFW distortion output from its ideally tracked value; the extra signal is effectively now compensating the compression of the EPR. At lower drive levels, however, the 9:1 IM backoff breaks down to a 3:1 slope and the corrective action is substantially impaired. The practical value of such nulls is debatable and will not be pursued; the possibility of their existence is, however, very important to note, especially when interpreting experimental data.

Figure 6.14 presents some more tracking data, which shows the effect of a nominal tracking error of 0.1 dB for three substantially different EPR selections. It is immediately clear from these plots that a higher EPR does not automatically improve the FFW system performance, other than perhaps in the highest power range; the backed-off performance is almost entirely a function of the tracking fidelity. Depending on the required performance

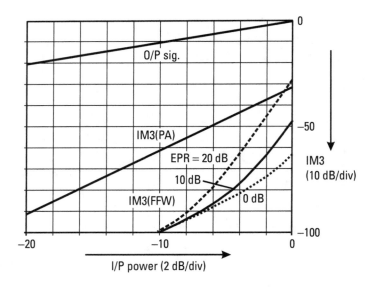

O/P sig.

IM3(PA)

EPR = 20 dB

10 dB

IM3(FFW)

0 dB

−50

−100

IM3
(10 dB/div)

−20   −10   0

I/P power (2 dB/div)

Figure 6.14 FFW loop IM3 performance with 0.1-dB tracking error, for three different EPR values.

goals, Figures 6.13 and 6.14 give the general impression that the major cost and efficiency impact of low EPR values (high EPA power) has a restricted performance payoff, as compared to excellence in tracking fidelity.

6.5.3 Compression Adjustment

In all of the PBO plots shown so far, it is clear that the EPA power capability is strained the most at the highest input drive level. This may seem an obvious statement; as the main PA is driven harder and into compression and AM-PM, the error signal reaching the EPA will increase. In the discussion surrounding (6.22), the concept of "compression adjustment" was introduced, whereby the output coupling factor α could be adjusted to cause the EPA input to be zero at any selected value of drive level. This is an issue of great importance in FFW loop design, and has been the subject of an ongoing debate in the FFW PA manufacturing and research community. It has some implications for adaptive FFW systems as well.

Equation (6.22) shows the mathematical possibility of selecting the α coupling coefficient such that the EPA input is zero at a chosen value of input signal amplitude v. More pragmatically, the selected value of α establishes a "gain standard" [3] which the overall loop action strives to maintain for all conditions of electrical, physical, and signal environments. Thus if a value of α is selected which corresponds to the inverse of the PA gain at its 1-dB compression point, an error signal will be generated at all drive levels *except* the 1-dB compression point, which will restore the overall gain to this predetermined level. In particular, at well backed-off levels, a more substantial error signal has to be generated by the EPA than in "normal" adjustment in order to reduce the small signal gain by 1 dB.

Figure 6.15 shows the effect of α selection on the FFW loop gain and the corresponding input power to the EPA. In order to analyze the entire FFW loop performance, the most generalized form for the EPA output once again has to be used, (6.23), along with the loop equation,

$$v_{of} = \gamma v_o + \beta v_{oe}$$

The resulting PBO characteristics make for an interesting comparison with the normally adjusted ($\alpha = 1/a_1$) results, shown also in Figure 6.15. As would be expected, the compression adjusted FFW loop distortion shows a deep null near the input drive level where the main PA gain corresponds to the α setting.[4] At the null point, and for a significant drive level range on either

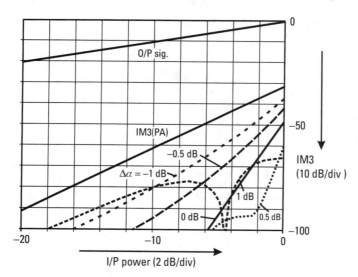

Figure 6.15 FFW loop IM3 performance with small changes in the first loop α parameter.

---

4. The compression adjustment causes a null in the fundamental signals reaching the EPA, but the IMs are still present and the exact position of the null is a more complex function.

side, the distortion is lower than the corresponding normally adjusted case using the same EPR. At lower drive levels, the compression adjustment cases show lower slopes and remain at higher levels than the rapidly vanishing characteristic in normal adjustment. Although this mid-range degradation of the compression adjusted FFW loop is a detraction, it must be recognized that the actual distortion levels in this region are very low and may fall within specification limits.

There appears to be a range of α values, corresponding to offsets of a few tenths of a decibel from the main PA small signal gain, where this effect appears to have an important potential for reducing the EPA power requirements. In making this assessment, it is important to realize that compression adjustment of the loop will result in less overall gain than the original PA, or the loop in normal adjustment. Consequently, when plotting output power and distortion characteristics, it is necessary to scale the input power sweep upwards by a factor of $1/\alpha$; this has been done in Figure 6.15, and all subsequent plots.

Examining the $\alpha = 0.5$ dB case more closely, Figure 6.16 shows a comparison between a normally adjusted FFW loop with an EPR of 10 dB and a 0.5-dB compression-adjusted FFW loop with an EPR of 15 dB. Clearly, over most of the upper 6 dB of power range, up to the 1-dB PEP compression

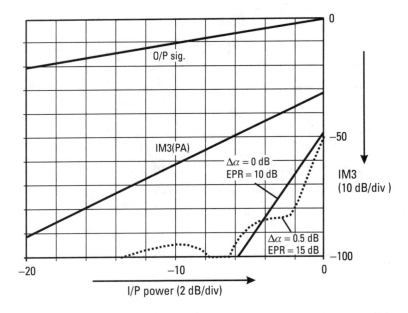

Figure 6.16 FFW loop IM3 performance with 0.5-dB compression adjustment and 5 dB lower EPA power, compared with normally adjusted loop.

point, the much more efficient combination actually gives lower IM3. In subsequent backoff to lower drive levels, the compression-adjusted loop distortion settles down to a 3:1 slope and in principle has increasingly higher distortion than the conventional configuration, although a 30-dB improvement over the basic PA is maintained. A kind of calibration can be used to assess the significance of the higher residual distortion in the compression-adjusted loop; a gain tracking error can be introduced into the normally adjusted combination. This is shown in Figure 6.17 and it can be seen that a tracking error of 0.1 dB is sufficient to bring the backed-off performance of the two systems to approximate parity. With this tracking error, the compression-adjusted loop, having an EPR 5 dB higher, gives lower IM3 distortion over the upper 9 dB of input drive range.

These results, although based on an idealized third-degree model, clearly demonstrate a mechanism by which comparable linearity improvement can be obtained, using substantially higher EPR and hence higher efficiency than in a conventional FFW system. Later simulation, using a more generalized PA model and a multicarrier signal, will show that the basic concept of compression adjustment as a means of reducing EPA power is robust and useful. In practical situations, the α adjustment will be part of the alignment process. It will frequently be found, on an empirical basis, to converge

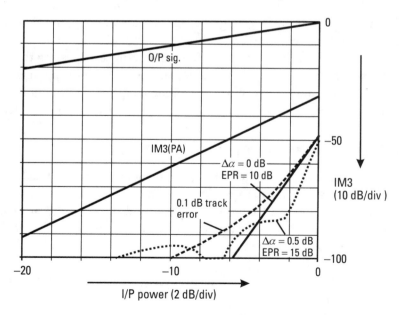

Figure 6.17 System of Figure 6.16, showing the "equalizing" effect of 0.1-dB tracking error on normally adjusted loop.

on an optimum setting which represents compression adjustment of the loop. For this reason, it has not received the theoretical recognition it probably deserves. The concept of compression adjustment also raises the possibility of dynamic adaption. In varying signal environments, compression adjustment could, in principle, be used selectively. This will be considered further in Section 6.8.

The above discussion on compression adjustment has so far not considered the inverse case of setting the α parameter to a value which represents a PA gain of greater than its linear value. Two such cases of this are shown in Figure 6.15, the decibel factor of α change now having a negative sign ($\alpha = -0.5$, $\alpha = -1.0$ in Figure 6.15). The immediate impression is that such adjustments cause rapid degradation in the corrective action of the FFW loop, due to the much larger signal which the EPA has to handle at all drive levels. Recalling, however, the discussions on error insertion coupling in Section 6.4, such α settings are now restoring the original PA output. In particular, the case of $\alpha = -0.5$ represents a system in which the transmission loss of the 10-dB error insertion coupler is restored by power generated by the EPA. Some caution is therefore required in dismissing the ensuing IM3 plot as being degraded in comparison to the normally adjusted ($\alpha = 0$) case; the entire FFW system is now able to supply the same nominal power as the PA itself, and the expense of larger power transistors to make up the output loss deficit is eliminated by selecting the appropriate value of α. The IM3 response could also be improved by using a lower EPR value. Although it could be argued that this is effectively the same thing as increasing the main PA power in terms of cost and efficiency, this approach does appear to have some possibilities, one of which will be considered further in Section 6.9. On the whole, however, compression adjustment seems to have more to offer.

6.5.4 Third-Degree Analysis: Conclusions

This section has used an idealized model to demonstrate the detailed operation of a feedforward loop in a quantitative fashion. The results serve to illustrate the complexity of a feedforward system, and the difficulty of making any generalizations about the various tradeoffs. The results of this analysis can be summarized as follows:

- A perfectly tracked FFW system shows a 9:1 backoff of third-order distortion products.

- Tracking errors result in a breakpoint from 9:1 down to 3:1 slope in the PBO characteristics.

- Distortion in the well backed-off region is primarily a function of tracking fidelity, not error PA (EPA) power rating.

- "Nulling" possibilities on the distortion output exist, both by using a positive tracking error to cancel EPA compression, and the use of compression adjustment to null the EPA input.

- The presence of two separate nulling mechanisms can lead to complex, but comparable, distortion suppression over a wide dynamic range with EPA power ratings substantially reduced from classical FFW configurations.

6.6 Feedforward Loop Simulation

The analysis in Section 6.5 was restricted to third-degree nonlinear effects, as far as the PA characteristics were concerned. This enabled quantitative analysis of some of the most important tradeoff areas in FFW loop design. It is reasonable, however, to question whether such conclusions can be relied upon, given some of the analysis and reasoning on PA models presented in Chapter 3. This section illustrates the impact of using higher-degree models which also contain AM-PM effects, through the use of an envelope simulator. It is still important, however, to define the models and to pursue the formulation up to a point where number crunching is left as the only option.

Initially, AM-PM effects can be ignored in order to demonstrate some important effects which will be observed in the PBO characteristics of the simulated FFW loop. If the PA output is defined to be the fifth-degree characteristic

$$v_o = a_1 v + a_3 v^3 + a_5 v^5$$

then the input to the EPA will be

$$v_e = v - \alpha \left\{ a_1 v + a_3 v^3 + a_5 v^5 \right\}$$

so for normal adjustment, where $\alpha = 1/a_1$, the EPA output will be

$$v_{eo} = b_1 \alpha \left\{ -a_3 v^3 - a_5 v^5 \right\} + b_3 \alpha^3 \left\{ -a_3 v^3 - a_5 v^5 \right\}^3$$
$$+ b_5 \alpha^5 \left\{ -a_3 v^3 - a_5 v^5 \right\}^5 \tag{6.30}$$

and in a perfectly tracked system the linear EPA output will cancel the distortion from the main PA, leaving a residual, uncorrected distortion term in the FFW loop output,

$$v_{fo} = \gamma a_1 v + \beta \left[b_3 \alpha^3 \left\{ -a_3 v^3 - a_5 v^5 \right\}^3 + b_5 \alpha^5 \left\{ -a_3 v^3 - a_5 v^5 \right\}^5 \right] \quad (6.31)$$

So the uncorrected output distortion contains a string of high-degree terms between the ninth and twenty-fifth. In the case of the third-degree analysis considered in Section 6.5, it was observed that this would mean first- and third-order distortion products would show a 9:1 backoff characteristic. In this case, the same conclusion applies to distortion products up to and including the ninth order, inasmuch as the lowest degree of distortion is still the ninth. For example, fifth-order intermodulation products will show a minimum of 9:1 backoff characteristic, and as higher degrees of distortion become more significant this slope will increase. This can come as a surprise when looking at simulated or measured results, since erroneous "inductive" reasoning might suggest that fifth-order products would show much higher PBO rates than 9:1, based on the more familiar jump from 3:1 to 9:1 displayed in a simple third-degree model.

Equation (6.31) also gives some indications that when higher-degree distortion effects are present, the possibilities for nulls in the backoff characteristics are much more numerous, due to increased number of terms which contribute to a specific spectral distortion product. This applies especially when compression adjustment is included, which adds a first-degree term to the EPA input.

Moving to a more generalized model for both PAs and an RF input signal $v\cos\omega t$, the main PA output becomes

$$v_o = a_1 v \cos \omega t + \frac{3}{4} a_3 v^3 \cos(\omega t + \varphi_3) + \frac{5}{8} a_5 v^5 \cos(\omega t + \varphi_5)$$

so the input to the EPA will be

$$v_e = v \cos \omega t - \alpha v_o$$

$$= v \cos \omega t - \alpha \left\{ \begin{array}{l} a_1 v \cos \omega t + \dfrac{3}{4} a_3 v^3 \cos(\omega t + \varphi_3) \\[2mm] + \dfrac{5}{8} a_5 v^5 \cos(\omega t + \varphi_5) \end{array} \right\}$$

and the EPA output is

$$
\begin{aligned}
v_{eo} = b_1 &\left[\begin{array}{l} (1 - \alpha a_1) v \cos \omega t + \dfrac{3}{4} \alpha a_3 v^3 \cos(\omega t + \varphi_3) \\[2mm] + \dfrac{5}{8} \alpha a_5 v^5 \cos(\omega t + \varphi_5) \end{array} \right] \\[3mm]
+ b_3 &\left[\begin{array}{l} (1 - \alpha a_1) v \cos(\omega t + \phi_3) + \dfrac{3}{4} \alpha a_3 v^3 \cos(\omega t + \varphi_3 + \phi_3) \\[2mm] + \dfrac{5}{8} \alpha a_5 v^5 \cos(\omega t + \varphi_5 + \phi_5) \end{array} \right]^3 \\[3mm]
+ b_5 &\left[\begin{array}{l} (1 - \alpha a_1) v \cos(\omega t + \phi_5) + \dfrac{3}{4} \alpha a_3 v^3 \cos(\omega t + \varphi_3 + \phi_5) \\[2mm] + \dfrac{5}{8} \alpha a_5 v^5 \cos(\omega t + \varphi_5 + \phi_5) \end{array} \right]^5
\end{aligned}
$$

$$(6.32)$$

and, as before, the FFW loop output is given by

$$
v_{fo} = \gamma v_o + \beta v_{eo} \tag{6.33}
$$

Equations (6.32) and (6.33), following further band limiting and simplification, enable the FFW loop output to be determined directly, for fixed values of v, and using envelope simulation techniques, for any chosen envelope domain time-varying function $v(\tau)$. This analysis is now most conveniently delegated to a math solver, or an envelope simulator. Initially, the system will be examined using a two-carrier signal,

$$
v = V \cos \Omega t \cos \omega t
$$

The goal here is to sample just a few results and conclusions which were obtained using third-degree simplification.

6.6.1 Effect of AM-PM in Main PA

Figure 6.18(a) shows a 20-dB input PBO sweep for two different PAs having the following normalized Volterra coefficients:

PA1 (with AM-PM): $a_1 = 1$, $a_3 = 0.2$, $\varphi_3 = 120°$, $a_5 = 0.1$, $\varphi_5 = 170°$

PA2 (no AM-PM): $a_1 = 1$, $a_3 = 0.1$, $\varphi_3 = 180°$, $a_5 = 0.05$, $\varphi_5 = 180°$

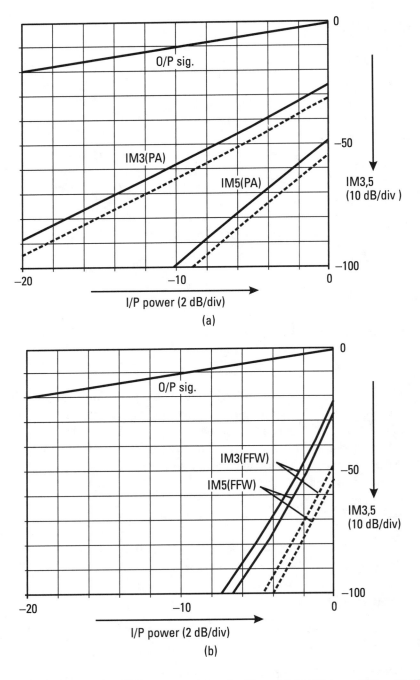

Figure 6.18 Two carrier IM responses: (a) main PA and (b) FFW loop, EPR = 10 dB; dashed lines represent PA having same gain compression but no AM-PM.

Note that both PAs have a 1-dB compression input voltage of unity, which represents the upper limit of the PEP power sweep in Figure 6.18. The IM3 sweeps show a slope which gradually increases from 3:1 towards 5:1 in the upper PBO range. The IM5 sweeps show a 5:1 slope due to the fifth-degree truncation of the Volterra series. Clearly, the effect of the AM-PM is evident on PA2.

Figure 6.18(b) shows the same pair of PAs placed in an FFW loop, which has an EPR of 10 dB and a 10-dB error coupler. The FFW loop is in normal adjustment (linear gain cancelled), and is assumed to be perfectly tracked. The EPA is assumed to have similar general characteristics to the main PA, with appropriate parameter scaling determined by the EPR selection.

The key observation here is the dramatic amplification of the detrimental effect of AM-PM in the corrective action of the FFW loop. Note also that the FFW IMs both show approximately a 10:1 slope due to the presence of nonlinear effects higher than the third degree.

Subsequent simulations will retain the AM-PM (PA1, solid traces in Figure 6.18), although this seems a clear target for improved PA design and device technology in FFW systems. The PA2 FFW plots can be compared directly with the third-degree analysis results in Section 6.5, and show similar IM3 levels. The detrimental effects of AM-PM in the PA2 FFW system result in little corrective action at the highest drive levels, and indicate the need either for a lower EPR value, or a backed-off PEP operation. In fact, compression adjustment will next be demonstrated as a possible relief in such drastic action.

6.6.2   Gain Compression Adjustment

The concept of compression adjustment has been discussed at some length in previous sections. The goal here is to check that the conclusions stand up to a test using a more real-life PA model. Figure 6.19 shows the effect of adjusting the α value to achieve fundamental cancellation at a point higher in the PBO range; in this case two values are shown, $\alpha = 0.25$ dB and 0.5 dB. The improvement in the FFW IM performance is quite dramatic, and compares quite well with the predictions using a simple third-degree model (e.g., Figure 6.16). Compression adjustment is clearly an important matter of practical adjustment in FFW systems.

6.6.3   EPR Change

Figure 6.20 shows the same system and conditions as Figure 6.19 but with the EPR value increased by a 5-dB factor to 15 dB. The IM levels in the

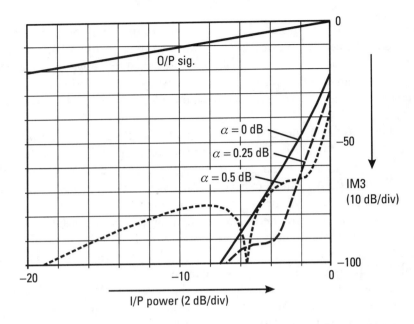

Figure 6.19 Simulation of compression adjustment; EPR = 10 dB (input sweep power scaled to allow for α value).

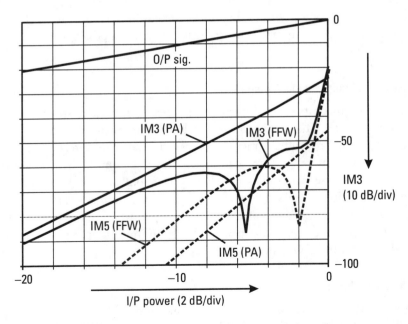

Figure 6.20 FFW system with 0.5-dB compression adjustment; EPR increased to 15 dB.

FFW IM response can be seen to increase by about 12 dB. Such a "flea-power" EPR could have useful applications in "budget" FFW systems, as discussed in Section 6.9.

6.6.4 Gain and Phase Tracking

Figure 6.21 shows the effect of a 0.5-dB gain and 5° tracking envelope on the normally adjusted system shown in Figure 6.18. The tracking errors are mainly detrimental, and show a tendency to restore the steeper IM slopes in the FFW output to their lower uncorrected levels. The nulling mechanism noted in Section 6.5 is still present.

6.6.5 Multicarrier Simulation

Figure 6.22 analyzes the same system as that in Figure 6.19, with the preferred 0.5-dB compression adjustment setting, and an eight-carrier input signal. The IM response shows the case where the eight-carrier signal has an equal PEP to the two-carrier signal shown in Figure 6.19. One important difference in a multicarrier signal environment is that the carriers cannot all be nulled by a single setting of the α parameter, as can be seen in Figure 6.22

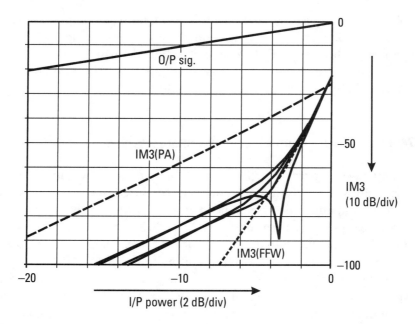

Figure 6.21 Effect of tracking errors (solid traces, ±0.25 dB, ±2.5°; IM3 shown only for clarity).

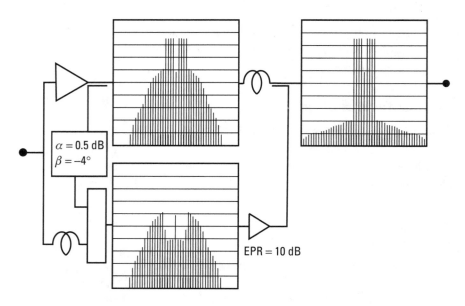

Figure 6.22 Multicarrier response of the FFW loop, showing partial nulling of fundamentals at EPR input.

by observing the spectrum at the EPR input. Clearly, a compromise value can be found, along with the companion phase standard setting, which gives a significant relief on the EPR requirement. But the basic observation that the optimum α value will change according to the signal environment has obvious implications for adaptive control of the FFW settings.

6.7   Feedforward Loop Efficiency Considerations

In almost any linearized PA application, *efficiency is king*. The primary reason for devising RFPA linearization systems in the first place is to reduce the power consumption which would be required to achieve a given power and linearity goal using a backed off Class A PA. A secondary consideration is cost efficiency, or *efficacy* in the best use of hardware. For example, a Class A PA operating near to its 1dB compression point may have an efficiency of 40%, an IM3 level of −20 dBc. If this is backed off 10 dB, the IM3 level will be approximately −40 dBc,[5] but the power will be 10 dB lower than the capability of the devices used in the amplifier, and the efficiency will be 4%. The key issue here, which seems occasionally to be missed, is that this is not the simple end of the story; in order to retrieve the original power level, either 10 such amplifiers need to be run in a power-combining scheme, or transistors with 10 times the current capacity have to be used, or some combination of these two options. So not only is there a tenfold reduction in efficiency, but there is also at least another tenfold increase in cost. A specification of −40 dBc is also still well short of most mobile communications requirements.

Clearly, even a basic FFW system can do a lot better than this, both in terms of linearity and efficiency. Although the final efficiency of a typical FFW system may appear low to the uninitiated, it needs to be pointed out that the standard of comparison is 1%, not the original 40%, based on comparable linearity performance. Two factors dominate the efficiency of an FFW system; the amount of PBO for a specified PEP level, and the power consumed by the EPA. The first of these is frequently more important than the second. The power consumed by the "housekeeping" functions may also need to be considered in a typical merchantable product. It is in fact a difficult and hazardous business to attempt to quantify the efficiency of an FFW system, taking account of such issues as tracking fidelity, PA modes, and device technology, not to mention the different signal formats and regulatory specifications. It seems therefore appropriate to consider a few of the basic factors which affect FFW system efficiency, without attempting piecemeal analysis.

It has already been stated that the final FFW loop efficiency will be dominated by the PBO-efficiency characteristic of the main PA itself. The widely perceived "low" efficiency of FFW amplifier systems is usually as much a function of the complex multicarrier signals which it is specified to handle, as of the extra power consumption of the EPA. The impact of EPA power consumption on overall efficiency is nevertheless worth quantifying, since it can be analyzed without making too many restrictive assumptions. This extra power drain is what sets the FFW loop apart from other linearization methods, and is seen as an Achilles' heel by detractors. In order to make a quantitative assessment of the impact of EPR power, some assumptions have to be made about both amplifiers. In the interests of efficiency, it is likely that the main PA will be of a "deep" Class AB design; this is certainly likely when using the beneficial IM nulling effects displayed by LDMOS devices. There is however an argument for using something closer to a Class A design for the EPA. Unlike the main PA, the EPA will typically not be

---

5. The only mitigating factor in this calculation is that in practice a well-designed Class A PA will probably show a significantly higher IM backoff than this (e.g., −50 dBc at the 10-dB backoff point, due to higher-degree nonlinearities only appearing near to the P1 dB level).

driven up to its 1-dB compression point, but will be much used at very well backed-off power levels. The more well-behaved and monotonic PBO behavior of a Class A PA will therefore be beneficial.

Operating close to Class B, the efficiency of a PA can be approximated to vary as the input signal voltage magnitude (see **RFPA**, Chapter 3). Assuming the main PA RF power output at 1-dB compression is P_{max}, then with an EPR (E), the EPA by definition has a 1-dB compression output power of P_{max}/E. So the backed-off efficiency of the main PA will be

$$\eta_{mpa} = \left(\frac{P}{P_{max}} \right)^{0.5} (0.65)$$

using typical P_{max} value of 65% for a deep class AB design. The EPA, being Class A, will have a typical efficiency of 40% at its own 1-dB compression point. A simplifying assumption can be made about the EPA; it will not significantly contribute to the main loop output power. As discussed in Section 6.4, this may not be quite true if the loop is used to correct power all the way up to the 1-dB compression point of the main PA. This, however, will have little impact in a dynamic signal situation, and so the EPA simply represents a constant extra current drain on the system.

So the dc power consumption of the two PAs at an output power P will be

$$dc_{mpa} = \frac{P_{max}}{(0.65)\left(\dfrac{P}{P_{max}} \right)^{0.5}}$$

$$dc_{epa} = \frac{P_{max}}{(0.4)E}$$

$$\eta_{ffw} = \frac{P_{max}}{\dfrac{P_{max}}{0.65\left(\dfrac{P}{P_{max}} \right)^{0.5}} + \left(\dfrac{P_{max}}{0.4E} \right)} \qquad (6.34)$$

which is shown plotted in Figure 6.23 for a range of EPR values and a 20-dB PBO range for v.

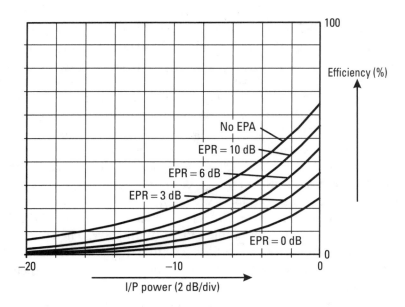

Figure 6.23 Efficiency plot for FFW loop, showing effect of EPR (see text for assumptions and definitions).

In the upper drive range, truly substantial intrusion of EPA power on the overall efficiency of the FFW loop is only seen for EPR values less than about 10 dB. A value of 10 dB would seem to be a good target for the EPR value, since higher values will pay diminishing returns in efficiency and undoubtedly stretch the error-correcting capability of the system. This conclusion is, however, modified by the consideration that a dynamic or multi-carrier signal will ride up and down the curve, and the time-averaged efficiency will correspond perhaps to somewhere around the 10-dB backoff point. At this lower drive level, a 10-dB EPR value drops the efficiency from 21% to 14%; this has a much more serious effect than the 65% to 56% drop at the PEP level, representing a 50% increase in dc supply power. These results do, however, represent something of a worst-case analysis, since they assume Class A operation for the EPA.

Equation (6.34) can also be used to assess another FFW efficiency tradeoff, that of PBO versus EPR. It is generally assumed that if a good tracking system can be implemented, then the 9:1 backoff slope of the distortion products can be utilized at the high end of the power range, and thus the EPA power specification can be reduced. This, of course, incurs the same device periphery issue as discussed at the start of this section; if a decision is made to use the PBO slope, the appropriate power level has to be restored to give the required rated PEP performance. Recalling (6.19), the relationship between b_3 and E is linear. E is, however, defined as a power ratio, so that third-degree distortion products from the EPA, which have voltage levels proportional to b_3, will appear on the output power PBO plot as having levels which vary as E^2. So if the main PA distortion backs off at a 9:1 rate on a decibel plot, a 1-dB PBO would correspond to a reduction in EPR of 4.5 dB to give the original distortion level at full drive. So if both PAs are scaled up by 1 dB, the overall distortion level will be the same as the original configuration, at the same drive level, but the EPR can be increased by 4.5 dB. Checking this using the plot shown in Figure 6.23, the 1-dB backoff point down the EPR = 6 dB curve shows an efficiency of about 45%; this jumps back up to an efficiency of about 52% on the EPR = 10 dB curve. The original efficiency, however, full drive on the EPR = 6 dB curve, was just under 49%. This small efficiency improvement comes at the very substantial expense of scaling the PAs.

Although this is only a single data point example, it emphasizes a different meaning of the term "PBO" in this context; we are really here talking about "periphery scale-up" (PSU), whereby for a given output PEP level, the main PA is set lower down its compression characteristic in order to reduce the IM distortion levels. This essentially means the replacement of the main PA with a new design using larger and more expensive transistors, and is not something that can be readily or quickly implemented in a given situation.[6] In fact, it is a matter of common knowledge and practice to use feedforward PAs somewhat backed-off from their 1-dB compression points at the PEP level. The decision as to what PSU factor to use is, additionally, a function of the signal environment itself. Signals such as those encountered in WCDMA systems, and multicarrier signals in general, have high peak-to-average ratios and the actual design value for "PEP" is less clear when the full statistical situation is analyzed.

The tradeoff between PSU and EPR, as a means of increasing efficiency, is a complicated function, which must include specific linearization goals for specific signal environments. In practice, an FFW system will have imperfect tracking and may use compression adjustment in the first loop, both of which will cause a sharp reduction in the 9:1 IM PBO slope. Cost also has to be included in any such analysis, since the use of PSU scales the

---

6. It should be noted, however, that small changes in PA power can be implemented by making tuning and/or bias changes, although such changes will themselves affect the efficiency and linearity of the PA.

transistor periphery of the whole amplifier chain. Quantitative analysis is therefore not attempted here. There is, however, one further avenue which is worth analysis, and that is the possible use of an envelope management system such as a Doherty PA.

The Doherty PA was discussed in detail in Chapter 2. Figure 6.24 shows an equivalent plot to that in Figure 6.23, but with a classical Doherty PA used as the main PA, rather than a Class AB design. To make a fair comparison, the "twin peak" efficiency of the DPA has been taken to be the same value (65%) as that used for the Class AB PA in Figure 6.23. The benefit of the DPA in the backed-off region is clearly seen, although the relative impact of the EPA on overall efficiency is similar in both cases. As discussed in Chapter 2, the DPA will probably, but not necessarily, show a slower IM backoff characteristic than a Class AB design having the same overall device periphery. Through the action of the FFW loop, this may be reduced to a negligible problem. The techniques described in Chapter 2 would appear to have an important part to play in improving the efficiency of FFW PA systems.

6.8 Adaption and Correction: Closing the Loop

A feedforward PA is fundamentally an open-loop correction system. The correction signal which is added to the main PA output at the output coupler is

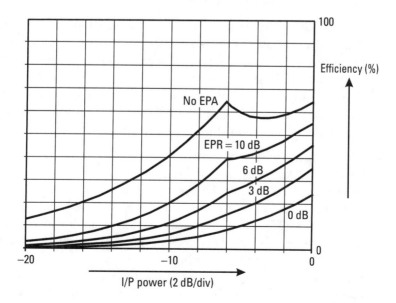

Figure 6.24 Efficiency plot for an FFW loop using a Doherty PA.

the final action of the linearization process; there are no further monitors or controls to determine, or improve, the adequacy of its corrective action. This is in stark contrast to any feedback system, where the corrective action of the gain and phase controls are under continuous iterative scrutiny by the feedback process. The benefit of the FFW system is that the inherent delay which a closed loop system introduces is essentially eliminated and the correction process can be performed at a much faster speed; the speed with which an FFW system can respond to a change in signal conditions is measured in terms of a few RF cycles. This speedy response to dynamic changes in PA gain and phase caused by nonlinear effects is the conceptual basis of the FFW system.

There are, however, other reasons why a PA gain and phase response may change. Temperature and aging certainly are two such reasons, but the overall malady can be more generally categorized as "drift." Drift in electronic systems seems to be almost as fundamental as the generation of noise, and can in a philosophical sense be linked to it. Any designer who is about to embark on a linearized PA product development which uses open-loop correction should perform the following experiment:

- *Equipment.* Your favorite RF PA (does not need to be high-power); RF network analyzer; signal source; two equal-power splitters; line stretcher; phase and gain trimmer; attenuator pads; and cables.

- *Setup.* As shown in Figure 6.25. The signal is split into two channels. The gain and phase trimmers are inserted into one channel, and the PA in the second channel, with enough output attenuation to neutralize (nearly) its gain. The two channels are recombined using the second splitter. A small value fixed attenuator and a line stretcher are placed in the second channel to balance the electrical length and gain of the PA in the other channel.

- *Measurement.* Sweep the network analyzer over about a 100-MHz bandwidth. Adjust the line stretcher and phase trimmer until a dip appears in the middle of the trace. Now adjust the gain trimmer to deepen the dip. Adjust the two trimmers successively and watch the null sharpen. With only a little care and attention, it will be possible to get a 20–30-dB null. Then the fun starts! Although it is clear that a deeper null is possible, getting it becomes increasingly difficult. Effects which were not noticeable in getting a 30-dB null suddenly dominate when 50–60 dB is attempted. For example, it becomes evident that the phase trimmer changes the attenuation slightly.

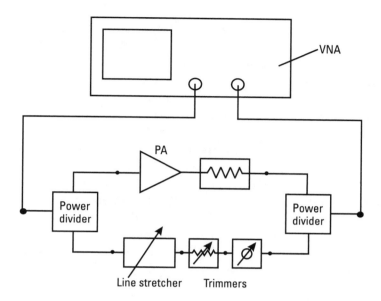

Figure 6.25　PA drift test.

Also, the gain trimmer changes the phase. There is a deep null somewhere in there, but "hitting" it seems now to be more of a trial and
error process, rather than a logical iterative one which it was to start
with. Got 50 dB? Now move your hands away, and watch it change.
O.K., still got it? Take a break; that was harder work than you
thought. Wander to the break room and have a cup of coffee. Good.
Back to the test bench … what's happened? The 50-dB null has
drifted off screen to another frequency, back to 30 dB now….

Performing this experiment is a good, and mandatory, experience in understanding the formidable nature of drift in an RFPA. There is a subsidiary
issue as well, concerning the ability to obtain a null over a substantial bandwidth. This raises issues concerning the phase linearity of the RFPA. A typical multistage RFPA will have an approximately linear phase versus
frequency characteristic, which will determine an optimum setting of the
compensating delay line. But there will also be a phase offset which is essentially constant with frequency. Such an offset can be easily converted into the
required 180° value for cancellation, using a phase trimmer. But to obtain
the required cancellation over a broader bandwidth requires some form of
all-pass network which gives a constant phase shift versus frequency.

Whatever the cause, drift is a formidable enemy in open-loop linearization systems, and over the years increasingly complicated methods have been devised in FFW systems to detect and correct for it [3–5]. This area has been a focus for patents in recent years, and any worker in the field is well advised to study the literally dozens of patented FFW adaptation and drift cancellation techniques before embarking on a commercial product development. The goal in this short section is to review the basic methods of adaption and drift compensation, without attempting to describe any specific implementation details.

The above experiment illustrates a basic problem which surfaces around the 30-dB cancellation point. The fact that drift problems are so readily observed at moderate levels of cancellation even in a simple system is highly disconcerting. In order to build a useful FFW system, drift compensation is a fundamental element. The requirements of a drift compensation scheme can be quantified, as shown in Figure 6.26. The 50-dB cancellation which was marginally achievable in the above experiment requires precision in the gain and phase adjustments of better than 0.01 dB and 0.5°, respectively.

Fortunately, as the trip to the break room illustrates, drift is a slow affair in comparison to the modulation speeds of typical communications signals. Indeed, picking up the time domain theme from Chapter 3, a fourth

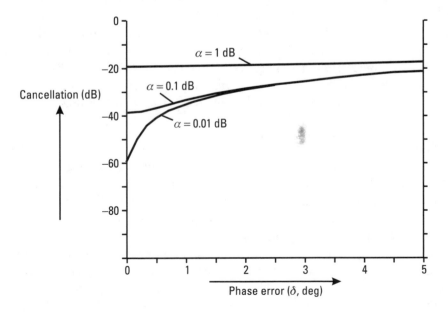

Figure 6.26 Cancellation errors in the FFW loop.

domain could be defined, the "drift domain," which is perhaps two or three orders of magnitude slower than the slowest of the three domains (RF, envelope, and measurement) already defined. It is therefore quite feasible to imagine a PA linearization scheme which uses feedforward to correct in the envelope time domain, but which uses a *quite separate* system to correct in the drift domain. Since the drift domain is so slow, feedback methods come right back into play. Indeed, since the measurement domain lies neatly between the two, it will be possible to use measurements as the basis for performing feedback compensation in the drift domain.

Figure 6.27 shows a conceptual implementation of a drift compensation scheme using amplitude envelope feedback. The feedforward system is assumed to operate with exquisite precision for a minute or two at a time. But drift slowly erodes the precision of the cancellation process in the FFW loop, and the IM products start to rise. This drift can in principle be sensed by the input and output detectors, in much the same way that was proposed in Chapter 4 as a method for actually linearizing the amplifier. Here, however, the video detection bandwidth will be reduced by many orders of magnitude so that the feedback loop only responds to very slow, long-term drift. It is in fact crucial that the integration time of the video detection circuitry is very long in comparison to the envelope domain variations of signal amplitude. The long integration time also has the effect of greatly magnifying the sensitivity to small errors.

It could be argued that a system such as that shown in Figure 6.27 is subject to further problems due to the drift inherent in the detectors and the

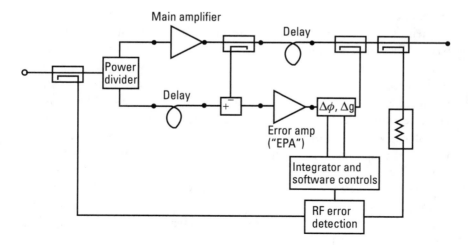

Figure 6.27 FFW loop with "drift domain" envelope feedback.

necessary post detection circuitry. At video frequencies, however, some of these problems have been already addressed by manufacturers of integrated circuits which are designed for such critical applications and themselves contain drift compensation functions. Simple envelope detectors will, of course, not respond directly to phase distortion in the output, and some refinements and additions would be required in order to implement a fully practical gain and phase tracking system. A refinement in the Figure 6.27 system would be to null the RF input and output signal samples, much in the manner of the first loop in the feedforward system. In principle, such a system will generate an error signal which can be due to either gain or phase errors in the FFW loop. On detection of an error signal, software routines can be used to determine which part of the system requires adjustment. The slowness of the drift process allows more than adequate amounts of time to send out test signals which probe various adjustment points in the system to determine which one reduces the error reading.

Although such systems have been developed into practical implementations, there remains a basic problem in the reliance on actual amplifier signals to act as test signals for calibrating the system. In particular, the detection process needs to be able to distinguish between errors caused by distortion in the EPA, and those caused by tracking errors. There is the additional problem of errors caused by drift in the detection process itself, the possibility for which increases as the complexity of the detection system increases. For these reasons the use of an internal calibration signal, or "pilot tone," has been found to be more satisfactory in many commercial FFW products.

Several decades ago, a few pioneers in the then fledgling field of radio astronomy discovered a simple method for greatly reducing the effect of drift in their amplifiers which were attempting to detect signals that were much weaker than the receiver noise level. The innovation, widely attributed to Dicke [6], is shown in Figure 6.28. A basic "total power" receiver attempts to detect a signal which may, for example, be 60 dB below the noise level of the receiver. The presence of this signal, following square-law detection and amplification, would represent a change in the noise level at the receiver output of 1 mV in a 1-V output. The source may take a number of minutes to traverse through the beam of the antenna, and during this time the receiver may well drift by many millivolts, so that on the output recorder the presence of the source is effectively submerged in the system drift. The Dicke receiver introduces a calibration source, which is conveniently a temperature stabilized cold load. The entire receiver is switched between the calibration source and the antenna, at a rate usually in the 100-Hz–1-kHz range. The key innovation was to have another switch, synchronized with the input switch,

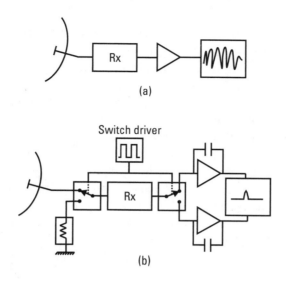

Figure 6.28 (a) Total power receiver and (b) Dicke receiver.

which sent the receiver output alternately to two similar video amplifiers, each supplied with suitably long time constant integration networks. Over a suitable period of integration, the amplifier which receives the antenna signal will show an easily measurable difference from the other channel, regardless of the system drift.

Such a system can today be recognized as a form of synchronous detector. It is well suited to the task of detecting very small variations in gain and phase tracking between two channels, so long as a suitable calibration signal and an appropriate "signature" are provided. Seidel [3] originally proposed a gain and phase tracking system for an FFW loop, based on the Dicke technique. Seidel's system, and many others since, inject a pilot test signal into the feedforward loop at the main PA output. The concept is then that such a "foreign" signal will be rejected and cancelled by the second loop, just as if it were a harmonically related distortion product. The key component is therefore the synchronous detector which samples the output signal. Even in the presence of high-power, multicarrier, modulated signals, the detector is able to pick out the signature of the pilot signal which is the indicator of gain or phase tracking errors. A modern commercial synchronous detector could obviously use a more complicated, and faster, signature than a simple 100-Hz squarewave. Seidel, in the 1960s era, had some problems in implementing suitable analog controls to perform the appropriate phase or gain corrections. In particular, figuring out whether the detected error was being caused either by a gain or a phase tracking error was in itself a considerable challenge. A modern implementation would be able to use software controls, along with optimization routines, to iterate any number of control elements around the loop to minimize the detected error signal.

Pilot carrier tracking systems are probably more reliable and robust than other kinds of tracking methods, but the presence of the pilot signal in the output spectrum of an amplifier is an unpopular feature with users, despite the fact that it may be 30 dB or so down from the main PA peak output power. More recently, the trend has been towards a higher level of internal test hardware, and a heavier reliance on software measurement and adjustment routines, as indicated in Figure 6.29. In particular, the availability of RFIC synthesizers and fast, high-resolution ADC converters enables the possibility of constructing built-in spectrum analysis at low cost, in comparison to the RF PA components. This enables the input and output spectra to be continuously monitored, and suitable software can be implemented which adjusts not only the gain and phase tracking controls, but can also minimize the EPA workload. Such a system seeks, in effect, to ship a complete test bench, along with a "virtual" test technician, with every product.

A final topic in this section is a revisit to the concept of compression adjustment that was analyzed in some detail in Section 6.5.3. It has already been shown, both in the third-degree PA analysis in Section 6.5, and confirmed in a more general simulation in Section 6.6, that under some circumstances, a slight under-setting of the output PA coupling coefficient, α, can

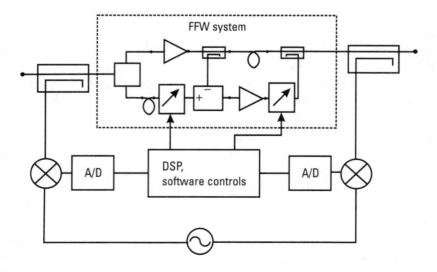

Figure 6.29 FFW system with built-in "virtual" test bench.

give some beneficial effects in an FFW loop, especially in regards to the EPA power requirement. It is therefore logical to speculate on whether a continuous adaptive control on this parameter may produce further benefits, under varying signal conditions. This is another much debated issue, and as such it is appropriate first to consider a *reductio ad absurdum* scenario. Suppose that the input signal is a two-carrier signal, thus being represented as a carrier which is vectorially modulated by a baseband sinewave. Suppose that an adaption system is designed which is fast enough to adjust the setting of α dynamically so that the input to the EPA is nulled out at all points in the modulated signal envelope. With such a system, the EPA receives no power at any point in the signal envelope; there is therefore no corrective action and one might as well throw away the feedforward loop components!

Such an argument emphasizes the dangers of taking the adaptive process too far. Adaption can only be performed over timescales that are slow enough that the parametric changes it implements do not themselves generate significant spectral components.

6.9 Variations

The FFW system lends itself to many variations. One such variation, the use of a predistorter on the main PA, has already been subjected to brief analytical treatment in Section 6.2. Unfortunately, it is such variations that form the basis for another litany of patents. Treading carefully around the minefield, two specific variations are worth considering in general terms: the double FFW loop and the extension of the FFW concept to the limiting case of an equal power combiner, where the main PA and EPA will always contribute essentially equal power to the output. A third "variation" is nothing more than a suggestion that simple basic FFW techniques can and should be used more extensively in conventional PA design.

6.9.1 The Double Feedforward Loop

The analysis and simulation results for the basic FFW loop presented so far indicate a system which, despite being a straightforward concept, will usually display complex behavior. Anyone who has actually attempted to adjust a real FFW PA on a test bench will have no difficulty in agreeing with the use of the term "complex," but would almost certainly wish to add "capricious" as a more apt characterization. With this background, it is hardly surprising that the mere suggestion of adding a second FFW correction loop around the first is met with suspicion, or even hostility, on the part of those who have to perform the final alignment and testing of a commercial product.

Notwithstanding this unpopularity, there seems some logic in favor of such a refinement, illustrated schematically in Figure 6.30. In particular, the second loop can be considered to be the means by which the first, or main, loop is closed and thus has its performance subject to further scrutiny and correction as in a feedback system. The downside, even overlooking the alignment problem, is the need for another error insertion coupler, delay line, and error amplifier ("EPA2") in the main PA output. With such detractions, the upside is less frequently considered. The following potential benefits would appear to be available:

- The power level of EPA2 can in principle be lower than EPA1, so the cost and efficiency impact may be almost negligible.

- EPA2 electrical length will, as a consequence, be much lower and possibly comparable to the insertion coupler itself.

- Theoretically, it can be shown that a perfectly tracked double FFW loop system has a 27:1 dB backoff rate of distortion products [3].

Possibly the core issue in all of the above considerations is the relationship between tracking and EPA2 power in the second, or outer, loop to the degree of linearization obtained in the first, or inner, loop.[7] Without entering into the complexities of a full analysis, it is clear that the correction process will have some different design criteria if the "main PA," which is now an FFW loop itself, has distortion products that are already at the −50-dBc level, rather than at −30 dBc. In particular, the linear cancellation process in the first loop of the outer FFW system will need to be much more precise than in the inner FFW system if the power level of the EPA2 is to be kept to a low value in comparison to EPA1.

This can be illustrated with a simple numerical example. If the inner loop is giving a signal amplitude of 100V with a distortion product at −40 dBc, the distortion will be at a level of 1V, which is then scaled down by the output coupling ratio ("α"). In the first loop of the outer FFW system, the main signal components at the input of EPA2 have to be cancelled, at least to the extent that the main signal component is much smaller than the dis-

---

7. The two separate, nested FFW loops are termed "inner" and "outer." The terms "first" and "second" loop apply as before to the error isolation and error correction loops which are within each FFW configuration.

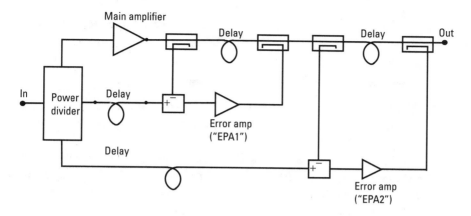

Figure 6.30 Double feedforward loop.

tortion component. With a system tracking capability which gives only, say, 20 dB of cancellation (e.g., 0.5 dB and 5°), there will be a residual uncancelled main signal component reaching the EPA of 10V, also scaled down by the α ratio. So in this example the EPA2 input is swamped by the imperfectly cancelled main signals. This does not, in principle, impair the linearization action of the outer loop, but it does place a much heavier demand on the EPA2 power level. Such a detraction will be evident in any FFW loop variation in which the main PA has been "improved" by any means; such would include the use of a predistorter, a signal linearizer-reconstruction technique (Polar Loop or Khan), or even just a well backed-off PA.

Clearly, the low levels of distortion emerging from the inner loop effectively force a much tighter tracking requirement on the first loop of the outer FFW system, or the EPA2 power requirement will be greatly increased. Even based on such simple numerical reasoning, it would appear that the original contention that the ratio EPA2/EPA1 can have a similar value to the inner loop EPR (EPA1/PA) can only be supported if very tight tracking is maintained in the outer loop. The degree of tracking required is closely linked to the distortion levels emerging from the inner loop. It can then be argued that it is simpler and easier to have such tighter tracking in the inner loop and dispense with the double loop concept. There would appear nevertheless to be some applications where the double loop approach may be the only option. Broader band systems, such as those encountered in satellite communications, or higher data rate wireless systems, have sufficient signal bandwidth that the gain and phase variation of PAs over the signal bandwidth itself are sufficient to cause substantial tracking errors. Such errors are much more difficult to correct using adaptive methods, and the double loop may be a better approach.

6.9.2 A Feedforward-Enhanced Power Combiner

The analysis and simulations in this chapter have shown that a feedforward system designed using an EPA which has much less power capability than the main PA, at least 10 dB lower, can reduce distortion products from the main PA by at least 30 dB in a well-tracked system. Such a system actually wastes the power capability of the EPA in terms of it ever contributing significant amounts of output power; the EPA function is entirely to linearize the main PA output. There is also a substantial reduction in the useable power output from the main PA due to the output insertion coupler, and other necessary components.

There is an alternative philosophy which can be applied to an FFW system. The concept is to use a much higher-power EPA but utilize its power to augment that of the main PA, in addition to performing some linearization functions [7]. The logical end-point of such a philosophy would be to use two PAs of equal power capability, and to combine their outputs using a "feedforward combiner," as shown in Figure 6.31. The system acts as a power combiner, so that the rated output power is approximately 3 dB higher than a single amplifier. But the "EPA" still additionally generates some error correction for the "main" PA, which gives a significant overall reduction in distortion levels in comparison to a conventional power-combined pair. Unfortunately, the EPA now has to handle much larger signals in order to generate its power contribution, and the uncorrected EPA distortion levels will be higher than in a conventional FFW system. Such a system nevertheless gives a potentially worthwhile improvement in the

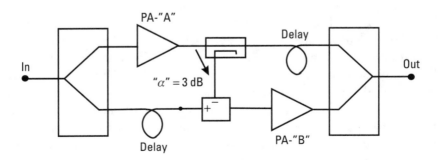

Figure 6.31 Feedforward power combiner.

distortion products at any power level, in comparison to the same PAs used in a conventional power combiner configuration, as shown in Figure 6.32.

The feedforward combiner still gives substantially higher distortion than that which can be theoretically obtained using an external feedforward loop around the combined pair, and an EPA at a −10-dB level. Nevertheless, it seems that if a power-combining scheme is being built in any case, the small additional cost of a FFW combiner may be worthwhile.

6.9.3 "Budget" FFW Systems

The feedforward system seems to suffer from an "expectation disadvantage," especially when being compared to other linearization systems. As the only viable system which can reduce nonlinear products by 30 dB or more, the extra component costs and power consumption tend to be viewed in a manner that is not always entirely fair or rational. On one hand, there is no doubt that multicarrier communications system specifications require very low IM levels, which represent at least 30 dB and maybe as much as 50-dB reduction from raw PA performance. Such performance requires a combination of at least two linearization methods. On the other hand, there are other

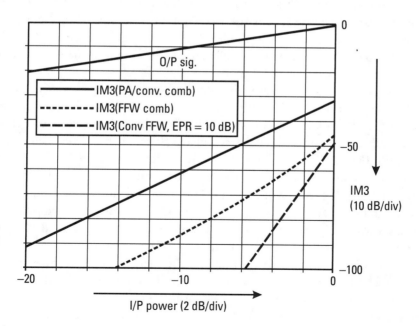

Figure 6.32 Feedforward power combiner performance, compared with conventional unlinearized combiner and conventional combiner with FFW linearization.

applications which have more modest requirements. In describing FFW systems, it somehow seems quaintly irrelevant to describe and analyze a system which gives, say, 10-dB improvement in IM performance. But a brief review of the literature of the last 10 years, especially the symposium circuit, will reveal many an enthusiastic researcher who proclaims a new predistortion or feedback variation which may give no more than 10-dB reduction in IM levels. Indeed, in many such cases the claimed improvements may be restricted to the low-order IMs and some higher order misbehavior may be carefully concealed by suitable choice of spectrum analyzer sweep range.

It is therefore relevant to speculate why the FFW technique has not become a more mainstream design practice; people recognize such techniques as push-pull, balanced, bias adaption, and power leveling as expected components of the PA designer's toolbox. All of these techniques require additional passive and active circuitry; they will substantially increase the size, cost, and complexity of the final circuit board assembly. A simple "budget" FFW loop could be incorporated as part of any PA design. The analysis in this chapter has shown that a system using an EPR of 10 dB can still give more than the stipulated 10 dB of IM reduction, with gain and phase tracking requirements that can be met using normal open-loop compensation techniques. Passive structures such as couplers and combiners are a familiar part of any PA layout and the requirements of coupling factors and directivity are not excessive if modest linearization goals are accepted. The use of a low-power EPA will typically reduce the length of the output delay line to quite manageable proportions.

Figure 6.33 shows the performance of such a "budget" system, based on realistic third- and fifth-degree models, including AM-PM effects, for both PAs. The simulation shows a 15-dB EPR value and the worst-case effect of a 1-dB and 5° gain and phase tracking envelope. Such a system can meet a −50-dBc IM specification for PBO levels greater than about 1 dB from the PEP P1dB compression point. The very low power EPA could be realized using a Class A design which would have minimal impact on overall efficiency and would give some further improvement on the simulated performance shown in Figure 6.33. The raw PA requires 12-dB PBO (and PSU) to meet the same performance specification. Realization of such a system could be done using a single board containing the main PA output stage and its driver. The EPA output stage would be a similar stage to the PA driver, and would require an additional low power gain block, which may well be at a power level where a low-cost RFIC could be used. The EPA delay compensation for such a low-power EPA could be realized using a suitable filter, which

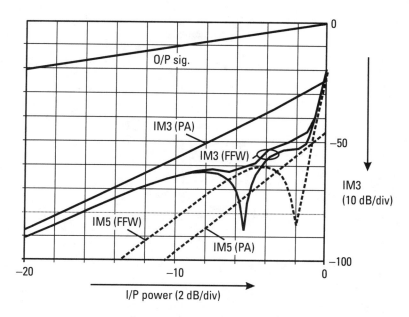

Figure 6.33 "Budget" FFW system: EPR = 15 dB, IM3 (solid traces) show worst-case effect of ±0.5 dB and ±2.5° tracking errors.

may be required as part of the design to eliminate harmonics and noise power. Both amplifier chains would need tight temperature compensation, but this is a likely specification requirement in any case.

6.10 Conclusions

The feedforward system is the mainstream technique for MCPA linearization. With suitable gain and phase tracking controls, it is capable of reducing the IM distortion of any PA by at least 30 dB, with no signal bandwidth or video delay restrictions. Even PA memory effects, displayed on a modulation domain timescale, can be fully corrected. In this sense the efficiency issue is not a direct detraction, since no other system has been demonstrated with comparable unrestricted linearization capability. It remains to be seen whether this supremacy will soon be challenged by improving DSP-based linearization systems which can potentially offer higher efficiency. The power of feedforward should not be ignored by PA designers with more modest linearization goals.

References

[1] Pothecary, N., *Feedforward Linear Power Amplifiers,* Norwood, MA: Artech House, 1999.

[2] Kennington, P., *High Linearity RF Amplifier Design,* Norwood, MA: Artech House, 2000.

[3] Seidel, H., "A Microwave Feedforward Experiment," *Bell Syst. Tech. J.,* November 1971, pp. 2879–2916.

[4] Cavers, J. K., "Adaption Behavior of a Feedforward Amplifier Linearizer," *IEEE Trans. on Vehicular Technology,* Vol. 44, No. 1, 1995, pp. 31–40.

[5] Narahshi, S., and T. Nojima, "Extremely Low Distortion Multi-Carrier Amplifier Self-Adjusting Feedforward (SAFF) Amplifier," *Proc. IEEE Intl. Comm. Conf.,* 1991, pp. 1485–1490.

[6] Dicke, R. H., "The Measurement of Thermal Radiation at Microwave Frequencies," *Rev. Sci. Instr.,* 1946, Vol. 17, pp. 268–275.

[7] Johnson, A. K., and R. Myer, "Linear Amplifier Combiner," *Proc. 27th IEEE Veh. Tech. Conf.,* Tampa, FL, June 1987, pp. 421–423.

7

Microwave Power Amplifiers

7.1 Introduction

Most of the present book, and also *RFPA*, has assumed that the primary application for PA design is in the worldwide mobile telephone network, which is presently confined to some narrow band allocations in the 830-MHz and 1,800-MHz regions. In fact, there are many other important applications for power amplifiers at higher frequencies, which can additionally require much larger instantaneous bandwidths than the mobile phone applications. The purpose of this chapter is to consider some of these, and describe some of the design and hardware differences they require. In particular, higher frequency and broader band applications will be discussed.

In consideration of other applications, it is appropriate to recall briefly the chronology of microwave technology. It seems possible to define three eras in the history of centimetric wavelength electronics. The first era, coinciding with World War II, was the heroic age of radar development. This was the period when the entire infrastructure of microwave technology was created, both in theory and hardware. Largely enshrined in the MIT Radiation Laboratory Series, the whole theory and practice of transmission, propagation, reception, and detection were thoroughly developed. This was the era of waveguides and vacuum microwave tubes, and extended up to at least 10 GHz in terms of practical devices and hardware.

The second era of microwave technology was more protracted, but can be readily linked with the post-World War II politics of the Cold War, and spanned those decades from the 1960s through to the end of the 1980s.

These decades of "military microwaves" resulted in many major new developments in the field, most notably the reduction in size caused by the application of new microwave semiconductor devices and high-density circuit layout techniques. This size reduction was not, until the very end of the period, achieved using integrated circuits as we would now understand the term. The circuit technology which enabled many of the military systems during this era was a hybrid one, using chip components attached to ceramic circuit substrates. This technology was so successful and ubiquitous that it became known as Microwave Integrated Circuit (MIC) technology, despite the fact that it was never anything close to a true monolithic IC process. Today, such technology survives, and is still a necessary part of making most microwave components and subsystems in the higher gigahertz bands. A review of MIC technology and techniques is the subject of Section 7.3.

The third era of microwave technology is, undoubtedly, the wireless communications era. It has been characterized by the need for high-volume, low-cost, yet complex systems and this has been enabled by the availability of true monolithic integrated circuits (RFICs). With more and more functionality built on to the chip and advances in high-frequency packaging, applications up to 2 GHz have been able to use enhanced, but otherwise conventional, PC board techniques to integrate circuits and systems. It seems likely, however, that in the future, and especially at higher microwave frequencies, an enhanced MIC system will be used. A basic understanding of the differences between "PCB" and "MIC" techniques is therefore a good base for approaching the requirements of future system generations.

It is only in the third era that broadband techniques have been sidelined. Virtually all military microwave systems are broadband in nature, frequently extending to multiple octaves. Much of the theoretical and technological developments in the Cold War military microwave era was focused on broadband techniques, and a whole industry developed during the 1980s for the manufacture of solid state amplifiers having octave or more bandwidth up to 40 GHz. The techniques and technology for these kinds of amplifiers will form another main subject for this chapter.

Alongside the military microwave developments, there emerged other important applications for microwave PAs, in satellite communications and microwave links. "Satcom" applications are usually characterized by moderate bandwidths, less than 1 octave, but wide enough to require broadband circuit design techniques. This application has generated another subcategory of PA design, using power devices which have been internally matched for specific satellite bands, usually a few hundred megahertz wide. The use of such products greatly eases the circuit design and supporting manufacturing technology, but nevertheless has its own quirks and limitations. These will also be discussed in this chapter.

7.2 Broadband Microwave Power Amplifier Design

7.2.1 Introduction

No discussion on broadband microwave power amplifiers would be complete, or even fair, without paying due respect to the traveling wave tube (TWT). This venerable device, now into its sixth decade, has changed remarkably little during that time, and it continues to perform microwave power amplification tasks that are still well beyond any conceivable solid-state device. Indeed, were it not for its single palpable weakness of short lifetime, the "billions and billions" of dollars spent on the development and manufacture of solid-state replacements may never have been spent. Work on TWT technology has, in fact, continued to the present day. The results of this work are never to be found splashed across the headlines, but a typical result [1] describes a TWT amplifier for the 6–18-GHz band, delivering 50W of power, with 30-dB gain at 25% efficiency. This tube itself is only 14-cm-long and about 1.5 cm in diameter.

The impact of newly available GaAs MESFET devices in the late 1970s and early 1980s was mainly to replace much lower power TWTs used for electronic countermeasure (ECM) receiver front ends. Such applications required high gain over octave bandwidths, but at power levels lower than 10 dBm. As MESFET power devices became available, this power was gradually increased, over similar octave bandwidths, up to about the 30-dBm level. In many applications, this power level was used to drive an output TWT. But in all cases, a radically different approach to circuit design and matching techniques was required. These techniques can be summarized under two headings: balanced amplifiers and network synthesis. These topics will therefore be considered first.

7.2.2 Broadband Matching Using Network Synthesis

At first sight, it would appear that the matching problem, which confronts the broadband microwave amplifier designer, can be thoroughly and completely solved by the conscientious application of the results of network synthesis, a branch of electrical engineering theory which dates back to the early part of the last century [2–4]. For a number of reasons which will be discussed, things do not turn out to be quite this straightforward; however, the theory can certainly be put to some use in these applications. Some simpler circuit element transformation theorems [5] form the basis of many of the matching topologies used in broadband solid-state design. A widely used example is shown in Figure 7.1. An inductive "tee" can be transformed into a mirror-image tee, having different values. The resulting input or output impedance level is then transformed by a factor n^2, where n is a function of the original tee elements:

$$n = 1 + \frac{L_s}{L_p}$$

$$L_{st} = \left(1 - \frac{1}{n}\right)L_p \qquad (7.1)$$

$$L_{pt} = \frac{L_p}{n}$$

The key aspect of this result is that the impedance transformation *is independent of frequency*; it is as if a piece of classical transformer theory has been retained despite the removal of any intended mutual coupling between the various inductors.[1]

The significance of this result is that it can be used, in essence, to transform a filter circuit into an impedance matching circuit. Figure 7.2(a) shows an example of this for the 4–8-GHz band. A filter circuit consisting of a cascaded series and shunt resonator can be designed either by using synthesis techniques or, more likely in the CAD era, by trial and error at the keyboard. The results of (7.1) can then be put to work on the inductors, giving the circuit in Figure 7.2(b). This circuit has exactly the same frequency response as that in Figure 7.2(a), but with an input source termination transformed down by a factor of $(1 + 1.25/2)^2$, or 2.64, a significant ratio for a frequency independent transformation. In a given application, however, it may not represent enough transformation; we are well aware that RF power transistors can require transformation ratios greater than 10:1.[2] This takes us back to the original filter and the need to increase the ratio L_s/L_p while maintaining adequate bandwidth. This latter requirement is more directly addressed by classical filter synthesis [3–5].

---

1. This transformation is the most useful case of a more generalized formulation, which can include capacitive elements.
2. It should be noted that broadband designs, with existing available RF device technology, will inevitably use lower-power devices.

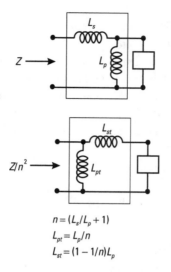

Figure 7.1 Inductor network transformations.

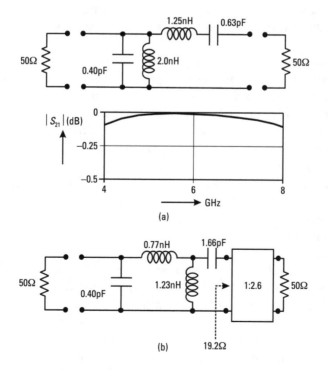

Figure 7.2 (a) Maximally flat 4–8-GHz filter and (b) transformation into matching network.

Before throwing this most venerable of kitchen sinks at the problem, it is worth using the CAD simulator to give some additional insight. Figure 7.3 shows the same network and subsequent source impedance transformation, but in this case higher Q-factors have been specified for the resonators; that is to say, their reactive impedances escalate more rapidly away from the resonant point over the specified frequency band. This results in a higher value for L_s and a lower value for L_p, and a correspondingly substantial increase in the transformation ratio

$$n^2 = \left(1 + \frac{L_s}{L_p}\right)^2$$

This increased impedance transformation has been achieved over the specified bandwidth without incurring any extra mismatch deviation within the band, referred to the respective transformed source resistance, compared with the original case shown in Figure 7.2. In fact, the two cases are close representatives of two classical filter types: Figure 7.2 is a Butterworth, or maximally flat, response, and Figure 7.3 is a Chebyshev, or equal-ripple, response.

On this initial analysis, it appears that the benefit of the Chebyshev filter for matching applications lies in its greater transformation potential. This is different from the classical filter design viewpoint, which cites the Chebyshev benefit as being the more rapid rolloff outside the specified ripple bandwidth. The well recognized disadvantage of the Chebyshev filter, its poor phase response, will, of course, carry straight over to the matching network equivalent, and this can be seen in the corresponding transmission phase characteristics of the two networks, shown in Figure 7.4. Although the difference in this particular example is certainly significant, it seems that in broadband designs the Chebyshev is an almost inevitable preference. A brief review of the specification and derivation of Chebyshev filter parameters is therefore appropriate.

Filter synthesis theory is a well-established and widely taught branch of electrical engineering, and is the subject of numerous dedicated books [2–4]. Broadband microwave amplifier design does, however, represent something of a peripheral application of the classical results. This is due to several factors, most notably the difficulty of realizing the ideal lumped elements at microwave frequencies, plus the issue already mentioned that matching, rather than filtering, is the main goal. Then there is the problem that the matching targets, usually the input and output impedances of RF transistors,

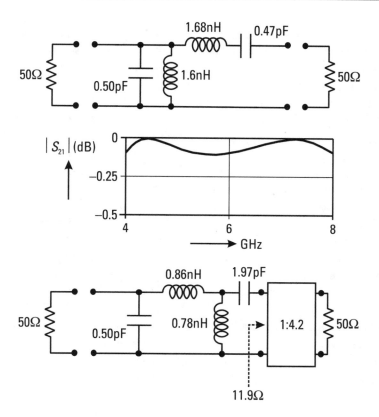

Figure 7.3 Equal-ripple 4–8-GHz filter and transformation.

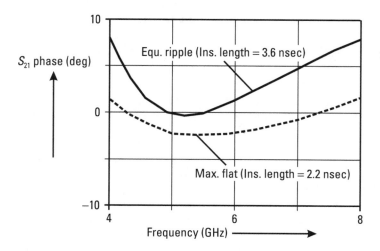

Figure 7.4 Transmission phase response of networks in Figures 7.2 and 7.3.

are not simple resistances. These factors restrict practical broadband micro-wave networks to the low "n" values, and the case of $n = 2$ will be the focus in this section. It will be seen that a wide range of broadband matching problems can be solved with $n = 2$ networks.

The classical starting point for designing a Chebyshev network is to establish values for the "prototype" lowpass network which give it a prescribed response. This is done by establishing the transmission response in the form of a polynomial in the normalized frequency Ω, and then by a successive comparison of the coefficients between the transmission response with the desired response, values for the elements can be obtained. Figure 7.5 shows a prototype network containing n elements, and Figure 7.6 shows the response of the corresponding nth degree transmission responses based on $C_n(\Omega)$, the Chebyshev polynomials. Clearly, a filter designer needs to select an appropriate value of n, based on the responses in Figure 7.6, in order to obtain the necessary filter characteristics. The selection of n for the design of a matching network is less clear; in this application the main concern is the in-band mismatch ripple, rather than the out-of-band attenuation. For the general case, the formal process of establishing the transfer characteristic of the network $An(\Omega, L_n, C_n)$, and the equating of successive coefficients of powers of Ω is lengthy and ultimately involves numerical iteration. The results are sufficiently well established that even dedicated textbooks often merely quote the results in the form of lengthy tabulations.[3]

Due to the specialized and restricted application we are considering here, it will be useful to deviate from the conventional generalized approach, and derive the element values for the specific but highly useful case of $n = 2$. Figure 7.7 shows the simple two-section lowpass prototype; a load resistor R_L has a series inductor L and a shunt capacitor C, forming a lowpass section. In order to write down a power transfer characteristic, it is also necessary to

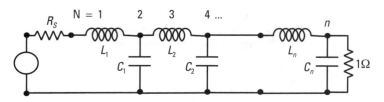

Figure 7.5 n-section lowpass prototype filter.

---

3. Many a hapless microwave designer has paged through these published tables, wishing that the de-normalization procedure had been more clearly specified!

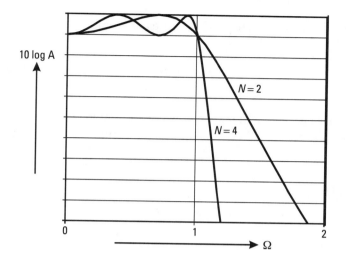

Figure 7.6 Responses of Chebychev lowpass prototype networks.

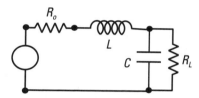

Figure 7.7 Two-section lowpass prototype.

specify the impedance environment in which the characteristic will be measured; this is defined to be a resistance of R_o. The transmission function of this network can be shown to be

$$A = \frac{1}{1-|\rho|^2} = \frac{1}{4R_o R_L} \left\{ \begin{array}{c} \omega^4 \left(LCR_o\right)^2 + \omega^2 \left(R_o^2 R_L^2 C^2 - 2LCR_o^2 + L^2\right) \\ + \left(R_L + R_o\right)^2 \end{array} \right\}$$

(7.2)

The transmission function of a second-order Chebyshev filter can be defined as

$$A = 1 + \varepsilon^2 C_2^2 \left(\Omega\right) = 1 + \varepsilon^2 \left(2\Omega^2 - 1\right)^2 = \omega^4 \left(4\varepsilon^2\right) + \omega^2 \left(-4\varepsilon^2\right) + \left(1 + \varepsilon^2\right)$$

(7.3)

where ε is a direct function of the in-band ripple, or maximum in-band mismatch,

$$\varepsilon = \sqrt{10^{R_p/10} - 1}$$

so that in more conventional microwave terminology, the value of ε is a design parameter related to the required maximum in-band mismatch ρ_{max}, where

$$R_p = \frac{1}{1 - \rho_{max}^2}$$

Comparing the Ω coefficients in (7.2) and (7.3) gives three relationships:

$$4\varepsilon^2 = \frac{L^2 C^2 R_o}{4R_L} \tag{7.4a}$$

$$-4\varepsilon^2 = \frac{R_o^2 R_L^2 C^2 - 2LCR_o^2 + L^2}{4R_o R_L} \tag{7.4b}$$

$$1 + \varepsilon^2 = \frac{(R_L + R_o)^2}{4R_o R_L} \tag{7.4c}$$

With some algebraic dexterity, these equations can be rearranged to determine the required element values L, C, and R_o, in terms of the single-design input parameter, ρ_{max}. The results are:

$$R_o = \frac{1 + \rho_{max}}{1 - \rho_{max}}$$

$$C = \frac{\sqrt{2(R_o - 1)}}{R_o} \tag{7.5}$$

$$L = \frac{2(R_o - 1)}{R_o C}$$

where the value of R_L has been normalized to unity. Note that this kind of network generates a specific value for the source load R_o; this is in fact a benefit to the main goal of impedance transformation, since from the first relationship it is clear that R_o will always be greater than unity.

At this point, we have the capability only to design the lowpass proto-type, which has an equal ripple bandwidth running from frequency $\Omega = 0$ to $\Omega = 1$. By a classical theorem [5], this lowpass characteristic can be transformed into any specified "real" bandwidth, using the network shown in Figure 7.8, with the following relationships:

$$R_{out} = \mathbf{R}_o R_{in}$$

$$L_s = \frac{\mathbf{L} R_{in}}{B}$$

$$C_s = \frac{1000 B}{\mathbf{L} R_{in} \omega_o^2} \qquad (7.6)$$

$$L_p = \frac{B R_{in}}{\mathbf{C} \omega_o^2}$$

$$C_p = \frac{1000 \mathbf{C}}{B R_{in}}$$

where for clarity the normalized prototype elements from (7.5) have been written in bold type.

The bandwidth, B, is defined as

$$B = 2\pi \left(F_H - F_L \right)$$

F_H and F_L being the upper and lower band edges in gigahertz, and the center frequency ω_o is defined to be

$$\omega_o = 2\pi \sqrt{F_H F_L}$$

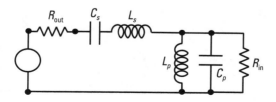

Figure 7.8 Two-section lowpass prototype transformed into bandpass filter.

this last definition being somewhat disconcerting to the RF designer in that the responses of these networks are not symmetrical about the center frequency in a linear sweep.

Equations (7.5) and (7.6) enable a four-element microwave matching network to be designed over any specified band, for a stipulated source termination and maximum VSWR (from ρ_{max}). The filter design is thus complete, but there are some further calculations to perform in order to determine the matching properties of the network thus designed. Referring back to the transformations shown in Figure 7.2, if the Norton transformation is applied to the inductors L_s and L_p in Figure 7.8, a new network, shown in Figure 7.9, results. The output impedance for this network has now increased by a factor of

$$\left(1+\frac{L_s}{L_p}\right)^2$$

giving an overall impedance transformation of

$$\frac{R_{out}}{R_{in}} = \mathbf{R}_o\left(1+\frac{L_s}{L_p}\right)^2$$

From (7.6), this can be written in the form

$$\frac{R_{out}}{R_{in}} = \mathbf{R}_o\left\{1+\frac{2(\mathbf{R}_o-1)}{\mathbf{R}_o}\left(\frac{\omega_o}{B}\right)^2\right\}^2 \tag{7.7}$$

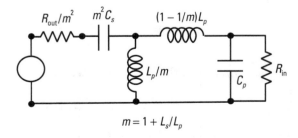

$$m = 1 + L_s/L_p$$

Figure 7.9 $N=2$ bandpass filter transformed into matching network.

This relationship shows that the transforming capability of the network is highly dependent on the fractional bandwidth of the network, represented by the ratio $\frac{\omega_o}{B}$.

It also gives a direct method of determining a value of $\mathbf{R}_o$ for a desired transformation ratio, a typical design situation. Equation (7.7) can be solved for $\mathbf{R}_o$, which leads directly to a value for ρ_{max}, thence to the element values given by (7.5) and (7.6).

Figure 7.10 shows a general set of plots of (7.7) and gives an overall picture of the broadband matching capability of $n = 2$ networks. One of the key differences between matching network design for active devices and passive filter design is the degree of tolerance in the ρ_{max} value. In amplifier design, a value of $\rho_{max} = 0.33$ corresponds to a 2:1 VSWR (9.5-dB return loss) and can be regarded as acceptable, representing only 0.5-dB reduction of potential gain from the device. Such a high value would be almost unmentionable in a filter design; this is not just a simple matter of sensitivities changing between the gain of an active device and the insertion loss of a passive component; the reciprocal nature of the passive element has the effect of doubling the effect of ρ_{max} in its insertion loss. With this level of tolerance on ρ_{max}, it can be seen from Figure 7.10 that the simple four-element network can perform some

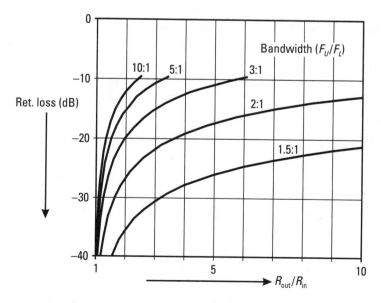

Figure 7.10 Four-element matching network performance summary.

quite remarkable impedance transformation tasks over very wide bandwidths.

The networks, as designed so far, still require some further work before they can be used in a practical design. In particular, the lumped element nature of the synthesized elements is not fully compatible with higher gigahertz techniques. Inductors, in particular, will usually have to be approximated by lengths of high impedance transmission line, and shunt capacitors will be realized using open-circuited shunt stubs (OCSSs). Series capacitors of low value represent a particular problem in that they cannot be easily substituted by a distributed element. A more fundamental issue is that the theory, as presented thus far, assumes that both source and matching target are resistive elements, at least over the frequency band of interest. This assumption can be modified if the reactive elements of the matching network are regarded as being part of the source or load. This introduces the general concept of "parasitic absorption," whereby the order of the network and the transformation ratio are selected such that the prescribed network input or output reactances match up with the parasitics of the target device.

The inclusion of parasitic absorption complicates and restricts the "exact" network synthesis procedure, whose inherent simplicity is one of its attractions. Although some admirable work has been published in this area [5], the power of the CAD optimizer has inevitably intruded, and ultimately sidelined an elegant *a priori* method. Nevertheless, experience suggests that a successful, or useful, broadband amplifier design will require the use of active devices whose parasitics are much less intrusive on the matching process than can typically be tolerated in high-power, narrowband PA designs. In the former situation, synthesis techniques can be used to obtain at least a viable topology, so that a CAD optimizer can be used to iterate the final element values, and incorporate both the device parasitics and the distributed element approximations. Clearly, at higher frequencies, the passive elements themselves will also have parasitics to be considered.

7.2.3 Balanced Amplifiers

The quadrature balanced amplifier was introduced, and discussed at some length, in *RFPA*. In broadband applications, use of the balanced configuration was mandatory when the MAG of the device was perhaps less than 6 dB, and was almost universal in commercially successful products of this type. Recalling the configuration, Figure 7.11 shows a pair of identical amplifiers placed between an input and an output quadrature 3-dB couplers. The great virtue of the balanced amplifier is that so long as the individual amplifiers are

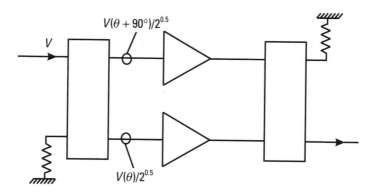

Figure 7.11 Quadrature balanced amplifier.

well matched, any mismatch reflections will be phase cancelled at the coupler input or output. This is particularly useful in octave, or greater, bandwidth designs where the matched gain of each device will display a slope of at least 6 dB/octave. Figure 7.12 illustrates this for a typical medium power 6–18-GHz design, based on a commercially available ½-W MESFET chip, the schematic for which is shown in Figure 7.13.

Although the gain response is fairly flat [Figure 7.12(b)], the accompanying return loss [Figure 7.12(a)] shows an input match which is reflective over the whole band. Any attempt to cascade such an amplifier would result in large gain variations due to the VSWR interactions. The balanced response, however, shows excellent match for basically the same gain response displayed by a single amplifier. Figure 7.12(c) shows another major advantage of the balanced configuration, a large increase in the stability or *k*-factor. This advantage seems to be largely ignored by designers of PAs for narrowband communications, the common wisdom being that the balanced configuration only displays advantages in wideband applications. For any bandwidth, however, the balanced configuration has another important advantage: It enables the use of a deliberate mismatch at each device whilst preserving a 50-Ohm termination for cascading purposes. The need for such device mismatch is frequently encountered in PA design, in working the various compromises between gain, power, efficiency, and linearity.

It should be noted that the mid-band coupling factor, for bandwidths in excess of an octave, has to be considerably tighter than the nominal 3-dB value. This results in a characteristic degradation to the VSWR response of the composite balanced amplifier. In practice, this VSWR response may dip close to 2:1 at the mid-band and band edge frequencies. Given that the

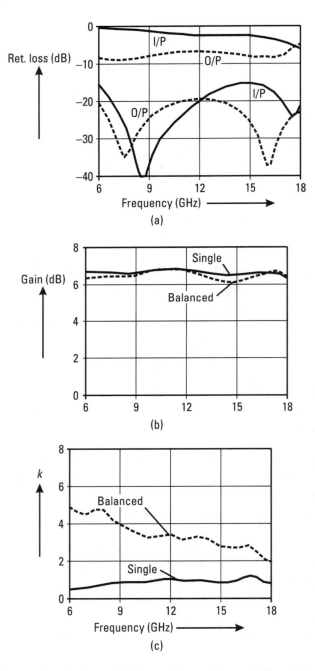

Figure 7.12 Schematic showing 6–18-GHz amplifier performance (simulation): (a) VSWR response, single-ended (upper traces), balanced (lower traces); (b) gain responses; and (c) *k*-factor responses.

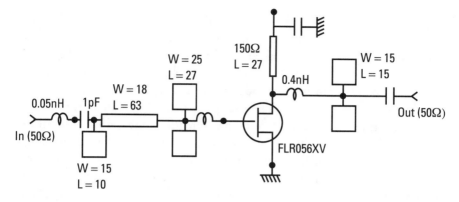

Figure 7.13 Schematic for 6–18-GHz medium-power (27-dBm) design. Design is based on 15-mil alumina; unmarked capacitors are assumed ideal blocking types at this stage of the design process.

driver and all of the previous gain stages may well display a similar VSWR response, the overall cascaded gain response can still show some VSWR dependent ripple. This is the penalty for using a simple quarter-wave coupling structure over such a large bandwidth, although as analyzed in *RFPA* (Chapter 10, p. 300) the overcoupling effect does not have a significant impact on the effectiveness of the power addition.

7.2.4 Broadband Power Amplifier Design Issues

Class A power amplifiers are commonly used in broader band higher frequency applications. The efficiency benefits of Class AB operation are harder to realize in these applications, due mainly to the harmonic termination requirements. Clearly, an octave bandwidth amplifier cannot present a load-line impedance over the whole band if the low end is also presented with a short circuit at the second harmonic. Even for bandwidths lower than a full octave, the design of a suitable network for broadband harmonic termination is problematic. In particular, it should be noted that in broadband applications, the likelihood of a transistor having a sufficiently large output capacitance to perform a harmonic termination function (see *RFPA*, p. 107) is very low. Such a large output parasitic would severely reduce the bandwidth capability of the device in the first place, and such devices will only find applications in narrowband designs.

The problem of designing broadband Class AB amplifiers which have comparable efficiency to their much narrower band counterparts remains essentially unresolved. There is a range of light Class AB operation which can give 5–10% efficiency improvement over a Class A design, but which is not significantly degraded by the absence of a harmonic short (*RFPA*, p. 107), but this seems to be a logical limit to the reduced conduction angle concept for broader band applications. One other avenue is, however, worthy of further discussion under this heading, the "push-pull" PA configuration.

A push-pull PA can in principle provide some of the benefits of reduced conduction angle mode operation over very wide bandwidths. Unfortunately, these benefits do not extend to the full efficiency which a conventional narrowband design would display, due to the same fundamental conflict of providing harmonic shorts simultaneously with broadband operation. The benefit of the push-pull PA is that it can, in principle, provide the same power management that reduced conduction angle modes provide; the fully driven efficiency will be lower, but the PBO performance will replicate a conventional narrowband design. Unfortunately, the second harmonic suppression which takes place at the output combining transformer of a push-pull RFPA does not in itself provide the same action as the harmonic short in a conventional single-ended Class AB design. This statement typically causes some consternation and is further illustrated in Figures 7.14 and

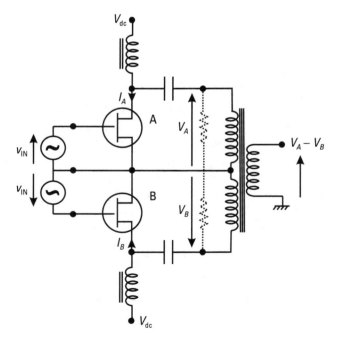

Figure 7.14 "Broadband" push-pull schematic.

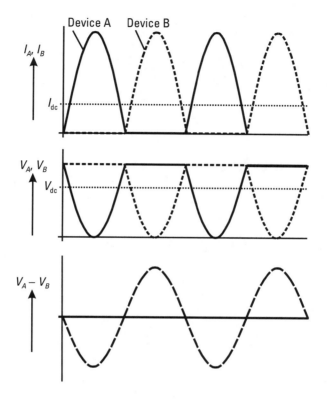

Figure 7.15 "Broadband" push-pull waveforms.

7.15. These show the basic elements of a push-pull PA with each individual device operating in an ideal Class B mode.

If it is assumed that the output center-tapped transformer behaves ideally over multiple harmonics of the fundamental input sinusoidal signal, the RF load presented to each individual device output will be a broadband resistance. With no harmonic short at either output, the voltage waveforms will be a simple resistive mirror of the rectified sinewave current waveforms. This is a significantly different situation from the classical narrowband single-ended case where these voltage waveforms would be sinusoidal, due to the action of a harmonic short. The non-sinusoidal voltage waveform has a detrimental effect, both on the available power and the efficiency at each device, because it has a lower fundamental component than a sinewave which has the same mean, or dc supply, value. The power and efficiency of a single device can be calculated, using the FET-style normalization and ideal transfer characteristic defined in Section 1.2.

For an input excitation having the form

$$v_{in} = v_q + v_s \cos \omega t$$

and setting $v_q = 0$ for Class B operation, the current for an ideal transconductive device will be

$$i_d = v_s I_{max} \cos \omega t \qquad -\frac{\pi}{4} < \omega t < \frac{\pi}{4}$$

$$= 0, \qquad \frac{-\pi}{2} < \omega t < \frac{-\pi}{4}, \frac{\pi}{4} < \omega t < \frac{\pi}{2}$$

The dc and fundamental components of current will be the same as for a classical Class B amplifier,

$$I_{dc} = v_s \left(\frac{I_{max}}{\pi} \right)$$

$$I_1 = v_s \left(\frac{I_{max}}{2} \right)$$

The output RF voltage has the form

$$v_o = V_{dc} + v_{pk} \left(\frac{1}{\pi} - \cos \omega t \right) \qquad -\frac{\pi}{4} < \omega t < \frac{\pi}{4}$$

$$v_o = V_{dc} + v_{pk} \left(\frac{1}{\pi} \right) \qquad \frac{-\pi}{2} < \omega t < \frac{-\pi}{4}, \frac{\pi}{4} < \omega t < \frac{\pi}{2}$$

where

$$v_{pk} = v_s I_{max} R_L$$

R_L being the broadband load resistance. At the maximum drive level, where $v_s = 1$, the load resistance has to be selected such that the RF output voltage just reaches zero at the point of maximum current, so that

$$v_{pk} = \left(\frac{V_{dc}}{1 - \frac{1}{\pi}} \right)$$

Note that this is a lower peak-to-peak RF voltage than in the harmonic short case, where the sinusoidal voltage swing would allow a maximum drive value of $2V_{dc}$ for v_{pk}. So the normalized RF output power at the fundamental will be

$$P_{rf} = \frac{1}{2}\left(\frac{v_{pk}}{2}\right)\left(\frac{I_{max}}{2}\right)$$

The "PUF" ratio, defined in **RFPA** as being the ratio of power to the Class A power obtained from the same device with the same dc supply, in this case reduces to

$$PUF = \frac{1}{2\left(1 - \frac{1}{\pi}\right)}$$

at the maximum drive level, which is −1.35 dB.

The efficiency at this maximum drive level is given by

$$\eta = \left(\frac{1}{2}\right)\left(\frac{V_{dc}}{2\left(1 - \frac{1}{\pi}\right)}\right)\left(\frac{I_{max}}{2}\right)\left(\frac{\pi}{V_{dc}I_{max}}\right)$$

$$= \frac{\pi}{8\left(1 - \frac{1}{\pi}\right)}$$

or about 57.6%.

Clearly, this is a lower efficiency than that which is obtained under the same conditions but with an output harmonic short. On the other hand, the PBO does show substantial "power management," as shown in Figure 7.17. The need to lower the value of the load resistance as the conduction angle is decreased, to avoid voltage clipping, actually results in a more optimum efficiency performance in the mid-AB region. The action of the output transformer is to take the two even harmonic-rich voltages from the two devices and add them together. The even harmonic voltage components will now cancel, giving a clean sinusoidal output, as shown in Figure 7.15 for the ideal Class B case. The waveforms for a mid Class AB case, shown in Figure 7.16, still show residual odd-degree distortion at the output. This is fundamental to Class AB operation, as discussed extensively in Chapter 1, and is not removed by push-pull operation. But in either case, the individual device voltages are not sinusoidal, which reduces the available power and efficiency.

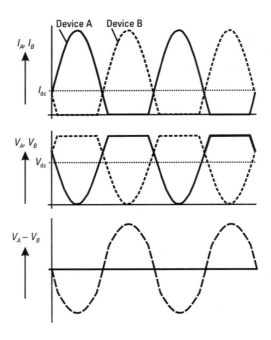

Figure 7.16 Class AB broadband push-pull waveforms.

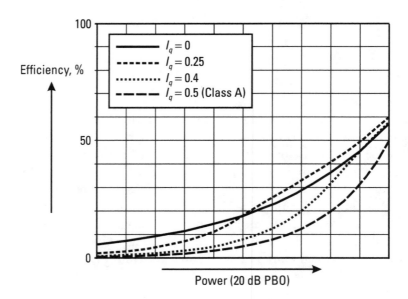

Figure 7.17 Efficiency/PBO performance for Class AB amplifier having broadband termination; load value adjusted to give unclipped voltage at maximum drive level.

In *RFPA,* it was shown (Chapter 4, p. 107) that the output capacitance of a typical RF power device may in some cases give sufficient harmonic decoupling to allow the efficiency to approach the ideal value. Unfortunately, such a device will be correspondingly difficult to match over broader bandwidths, and this approach can only be expected to give a modest improvement over the performance figures obtained above, which assume a resistive load at each device.

As discussed in *RFPA,* the main advantage of push-pull operation to the microwave PA designer is the series combining of two low-impedance RF power transistors. This gives a 4:1 advantage over parallel combining of the same devices. Even this advantage, however, is compromised by the need for a broadband balun transformer, to serve the function of the center-tapped ideal transformer shown in textbook schematics such as Figure 7.14. Although balun devices are fairly straightforward to implement for narrowband applications, octave or higher bandwidths at gigahertz frequencies pose much greater problems. The basic "bazooka" structure, shown in Figure 7.18, can possibly be stretched to an octave bandwidth, the bandwidth essentially being a function of the highest practical value the impedance Z_B can be made. But the use of a physical piece of cable becomes impractical at higher frequencies, and a printed planar equivalent is needed. Such printed structures have a frustrating habit of never working quite as well as the coaxial prototype. The ongoing search for a compact, planar, broadband balun structure can be followed in the annual offerings in symposium proceedings and patent abstracts, and is not reproduced in detail here. Typical claims for new balun designs frequently do not pass muster in PA applications, where both losses and asymmetry in the coupling responses can significantly degrade the overall performance. This unfortunately eliminates many structures, which may satisfy the needs of mixer designers admirably, from consideration for PA use.

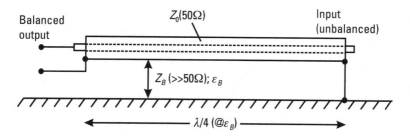

Figure 7.18 Coaxial balun structure.

7.3   Microwave Circuits and MIC Techniques

7.3.1   Introduction

At frequencies higher than about 2 GHz, the parasitic reactances associated with packaged components become a rapidly escalating limitation, especially if broad bandwidth performance is required. Although this problem can be, and still is, reduced by the development of better packages, the more radical solution of eliminating the package altogether has been the basis for most broadband microwave circuit design for many years. Such circuit technology has been variously called "MIC," "chip-and-wire," and "thin film hybrid," but all these terms refer essentially to a circuit technology containing the principal ingredients shown in Figure 7.19. A ceramic substrate, usually having a thickness in the range of 0.25–0.65 mm (10–25 mils), has a vacuum-deposited refractory metal coating. Using a combination of plate-up and etch-down masking operations, a circuit pattern is defined. The metallization will usually include a layer having a prescribed sheet resistance, typically around 50Ω/square, enabling surface resistors having very low parasitic capacitance to be formed along with the metal circuit pattern. Plated-through vias may also be used for grounding purposes.

The chip components themselves are directly die attached to the substrate, usually using small preforms of an alloy such as Gold/Tin (AuSn). The attachment has to be performed in an inert atmosphere at a temperature in the vicinity of 300°C. Traditionally, this die attach operation was performed by hand and required considerable skill. Power transistors, in particular, require a certain amount of "scrubbing" in order to get satisfactory heat conduction, and this highly delicate operation was usually not helped if the operator was made aware of the high cost of the chip. A typical sub-assembly would require the attachment of many individual chips, and so the entire operation had to be performed at some speed, since the many metal layers involved in a die attach operation will start to diffuse and deteriorate in a matter of a few seconds at the high temperature required for the operation. It is probable that some of the high-volume machines and processes which have been developed for modern RF PC boards could be modified to handle an

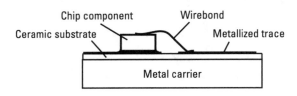

Figure 7.19   "MIC" circuit elements.

MIC process. The availability of good quality conductive epoxy has also eased the die attach requirements. Once attached, the chip components usually require some bondwire interconnects down to the substrate. Traditionally, these were done using 20 micron gold wires and thermocompression bonding; both the chip contacts and the circuit traces were gold. This process required the substrate to be heated, but to a lower temperature than that required for die attach.

Largely due to the difficulties of manual die attachment of many individual chips, the MIC substrates were often limited to single functions. These MIC "modules" would be built to standard sizes, dimension options and pinout positions being rigorously stipulated. For example, a microwave amplifier may well be designed using one module per stage. This modular approach has many advantages, in terms of multiple use of designs and manufacturing efficacy. It does, however, raise the problem that the thin ceramic substrates have to be mounted on to some form of metal carrier, which needs to have well matched thermal and expansion characteristics to the ceramic substrate. Nickel-Iron alloys which have well matched thermal expansion coefficients to glass and ceramic have been widely available since the tube era, but in general have poor thermal conductivity. Molybdenum is an obvious compromise candidate (see Figure 7.21), but is a difficult material to work with, both in terms of its machining and plating properties. In more recent times, some special materials have been developed, which consist of a tungsten matrix impregnated with copper. Although not widely available, these copper-tungsten composites have become the favored material for MIC substrate carriers.

One of the principal disadvantages of MIC technology is the requirement for a hermetic enclosure. Bare semiconductor chips and their associated bondwires simply cannot be housed in normal atmospheric conditions. This would appear to be a serious problem for possible future commercial use of this technology, although the food industry seems to have tackled it with admirable effectiveness. Molecules are a lot smaller than bugs, but the hiss of air when opening a canned product has been a domestic quality assurance guideline for long enough to indicate that hermetic sealing need not be a prohibitive manufacturing process. The need for low-loss, high-quality microwave transitions through a hermetic housing is an additional problem.

7.3.2 Substrate and Heatsink Materials

The most common substrate material for MIC work is alumina, the amorphous form of aluminum dioxide. It has a low dielectric loss tangent, such that for small lengths of high gigahertz transmission line the losses are dominated by the resistive loss in the traces. In the early days of MIC technology alumina substrates had to be polished in order to obtain satisfactory microwave loss performance; however, this process greatly increased the difficulty of metal deposition. In more recent times, "as-fired" alumina substrate blanks with adequate surface flatness for MIC circuit work have become readily available. The dielectric constant of just under 10 allows for very compact circuit layouts.

Being a ceramic, there is a thermal conductivity issue. If power chips, such as RFPA devices, are being used in an MIC circuit design, there are two basic approaches to tackle the heat dissipation requirements. The most popular approach is to use a metal rib, or septum, in order to provide a good thermal sink and also to reduce the ground inductance in comparison to a plated via arrangement. Such a configuration is shown in Figure 7.20. An alternative approach is to use a substrate material with a higher thermal conductivity. Aluminum Nitride (AlN) substrates are now available and have the same dielectric constant as alumina. But AlN only offers an improvement factor of between 3 to 5 over alumina, which is a long way short of the improvement offered by a metal septum (see Figure 7.21). Beryllium Oxide, BeO, has gone well out of favor, and indeed is not allowed in many applications due to toxicity problems. The chart of material properties, Figure 7.21, shows that the use of copper as a carrier or heatsink septum has to be carefully considered due to the much lower thermal expansion rate of both silicon and gallium arsenide die. Kovar (Kv) and Alloy 42, both commonly used as carrier materials for their alleged thermal match to alumina, can be seen to have desperately poor thermal conductivity, lower even than alumina itself.

Diamond heatsinks have played a role in microwave power technology for several decades, especially in the era of Gunn and Impatt diode oscillators. Natural diamonds of the type IIA category have no use as gemstones and are a relatively cheap byproduct of the diamond mining industry. Such materials have by far the highest thermal conductivity of any other known material, as much as 5 to 10 times higher than copper. More recently,

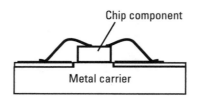

Figure 7.20 MIC configuration for higher dissipation components.

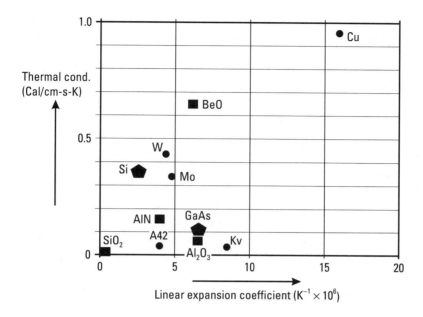

Figure 7.21 MIC material properties.

artificially synthesized diamonds have become available for electronic heatsinking applications, but typically do not display such spectacular performance in comparison to copper. Unfortunately, diamond is not an electrical conductor and thus poses RF grounding problems if used as a heatsink for most RF devices, which use the underside die attach area as the common lead connection.

7.3.3 MIC Components and Structures

MIC technology offers much higher precision and resolution in terms of line widths and gaps, in comparison to normal PCB copper foil techniques; a typical process can offer line and gap resolutions of around 5 microns. It also offers the ability to "print" resistors of user-prescribed dimensions as part of the metallization process. The core structure, however, is the transmission line, and there is an immediate choice here between two quite disparate schools of thought: open microstrip and coplanar waveguide, shown in Figure 7.22. Microstrip offers simplicity of layout and a closer relationship with more conventional PC board techniques. On the other hand, it suffers from radiation effects which can escalate into major stability and resonance

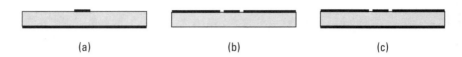

(a)   (b)   (c)

Figure 7.22 MIC transmission lines: (a) microstrip, (b) coplanar waveguide (CPW), and (c) CPW with ground plane.

problems when multiple gain stages are placed into a metal enclosure. There is also a problem in grounding shunt elements, which basically requires plated vias or metallized wraparounds at the substrate edges. Coplanar waveguide has been touted by a vociferous minority, who claim that this structure neatly solves both of the above problems. It does, however, have some problems of its own. Most notably, the need to space the substrate well above the ground plane formed by the metal enclosure can be inconvenient; coplanar waveguide starts to lose some of its advantages when fabricated using a substrate with a lower ground plane [Figure 7.22(c)]. There are also various moding and transition issues which result in the need for multitudes of wirebonds which interconnect the ground planes on either side of the transmission lines.

One component structure which is conveniently easy to realize in microstrip MIC is the directional coupler. The higher resolution of the MIC metallization process enables fine gaps to be specified, and coupling structures down to about 6 dB can be easily manufactured *in situ* without any need for extra drop-in components. Possibly one of the most important innovations in the military microwave era was the interdigitated, or Lange, directional coupler [6]. This structure multiplies the coupling between two adjacent microstrip transmission lines by splitting each line longitudinally into two or more sections (Figure 7.23). Coupling factors as low as 1.5 dB can be realized using this structure, depending on the ultimate resolution of the process in use. Octave band 3-dB couplers typically require gaps of about 1 mil (25 microns) between individual interdigitated line sections that are between 1 and 2 mils wide; these dimensions would apply to a quarter-wave structure having four interdigitated "fingers." Most modern linear microwave CAD packages include a measurement-based model for the design of such couplers, and in practice they are fairly tolerant of process variations. The basic layout shown in Figure 7.23 can pose problems for wirebond placement, and enlarged pads for accommodating these have been found in practice to have negligible effect on the predicted coupler performance. All microstrip couplers suffer from reduced directivity due to the mixed dielectric environment. Two well-known tricks are available to improve the

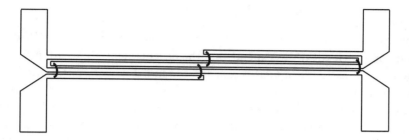

Figure 7.23 Lange coupler.

directivity in applications which require it, and are shown in Figure 7.24. Small capacitors can be placed across the coupler to connect opposite ports, or alternatively the gap can be "wiggled" in order to increase the electrical length associated with the air-dominated odd-mode.

Although the MIC designer will strive to use transmission line matching networks as much as possible, there will always be requirements for inductors and capacitors, as a minimum in bias insertion networks. In the

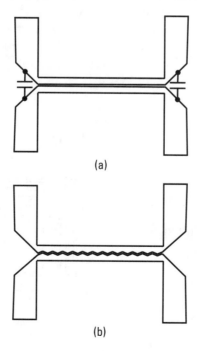

(a)

(b)

Figure 7.24 Techniques for directivity improvement in microstrip couplers: (a) shunt capacitors and (b) "wiggled" gap.

2–8-GHz range, even the matching elements can be quite large if realized entirely using transmission lines and stubs, and in the early era of MIC development, much effort was expended in the development of quasi-lumped elements such as spiral inductors and interdigitated capacitors (Figure 7.25). Such components tend to stretch the resolution requirements of the metallization process and have Q-factor limitations. The interdigitated capacitor has been largely superseded by smaller chip capacitors or by multilayer structures. The spiral inductor has become something of an icon for the MMIC industry and lives on, for better or for worse.

An interesting and more recent development has been several advanced MIC processes which use multilayer structures [7]. These processes can offer both air bridges, to eliminate the need for bondwire interconnects, and integrated metal-insulator-metal (MIM) capacitors. In many respects, and in actuality in some cases, these processes were a by-product of monolithic microwave IC (MMIC) development. The availability of higher-quality dielectric substrates enabled semiconductor processes and equipment to be used for MIC applications. One problem with these processes was the difficulty of making ground vias in the much thicker dielectric substrates, in comparison to the semiconductor wafers. Although touted as future high-volume, lower-cost alternatives to RFIC processes, the RFIC seems to have emerged as the preferred technology for current commercial applications up to 2 GHz. The advanced MIC processes may still have a part to play in future higher-frequency applications.

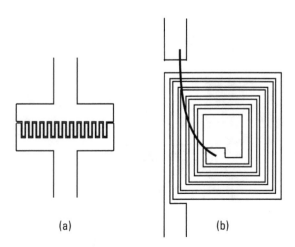

Figure 7.25 "Lumped" element MIC elements: (a) interdigitated capacitor and (b) spiral inductor.

7.4 PA Design Using Prematched Modules

7.4.1 Introduction

Despite the voluminous literature on PA design, it is an irony that a majority of current higher-power PA products are built around pre-matched modules. These products are power transistors which have matching networks built into the package, so that they interface directly with a 50-Ohm environment. The "designer" is thus freed from the need to design matching networks. Perhaps more significantly, the PA manufacturer is freed from the responsibility to handle the small changes in device impedance characteristics which will usually not be guaranteed by the manufacturer for the discrete, unmatched, equivalent part. Internally matched transistors are usually available as catalog items for the established satellite communications, link, and mobile communications bands. Their performance is guaranteed and 100% tested by the manufacturer prior to shipment. The attractions of this concept are obvious for a production operation. There are, nevertheless, some disadvantages and certainly some issues associated with the supporting microwave circuitry that are worth some quantitative analysis.

7.4.2 Matching Issues for IMTs

Internally matched microwave transistors (IMTs) are mainly used for the highest power stages in a PA. At the present state of RF power technology, the use of IMTs becomes quite common in the gigahertz range above a few watts of power, and almost mandatory in the 10–100-W range. This is not a simple matter of a lack of qualified microwave designers to design the necessary matching networks around discrete devices. Basically, once the power gets up to these higher levels, and especially at higher gigahertz frequencies, the transistor matching problem develops into a three-dimensional electromagnetic nightmare.

This is due in part to the very low impedance levels, but equally dependent on a number of unwelcome novel effects which can usually be ignored at lower power levels. In particular, the effect of common lead parasitics can escalate to alarming proportions with large periphery devices and can be frustratingly difficult to reduce. Another major issue is the integrity and uniformity of the heat removal across the very large semiconductor dies.

The high impedance transformation ratios require almost outrageous values of inductances and capacitances which simply cannot be realized in discrete packaged forms; the inductors in particular have to be made using many parallel wirebonds whose precision and uniformity can only be achieved using specialized automatic machines (or still, in extreme cases, specialized non-automatic human skills).

Although these various challenges and difficulties have nominally been solved by the manufacturer of an IMT, it is still necessary for the end user to be aware of the sensitivity of the internal matching networks to the external impedance environment. The concept of a "50-Ohm" environment has to be quantitatively redefined in a more precise manner than is typically necessary in RF design. This is illustrated for a specific case in Figure 7.26. An IMT output matching example from *RFPA* has been used. Details of the design of the matching network will not be repeated here; a device which would be typical of a 20-W LDMOS type is shown, requiring a loadline impedance of 12.5Ω. This very moderate impedance is dramatically transformed down to about 1Ω by the device output capacitance. We assume, however, that the matching elements are appropriately realized, along with an input-matching network, in a neatly sealed package. The issue now arises as to how sensitive such a device will be to an imperfect 50-Ω termination; indeed, there is a subsidiary question as to what precise termination, both in magnitude and phase, the manufacturer used when aligning and testing the device prior to shipment.

Figure 7.27 shows the detrimental effects of terminating such a device with imperfect loads, using the assumption that it was originally set up with a perfect termination. Using the loadline theory (*RFPA*, Chapter 2) it is possible to map the full phase contours at chosen VSWR levels onto the load pull contours for the device. It can be seen that even a 1.2:1 VSWR (20-dB return loss) load can "pull" the carefully optimized impedance to reach the 1-dB loadpull boundary. A more common VSWR level, representing a 13-dB return loss, has much more serious consequences. These simulations are

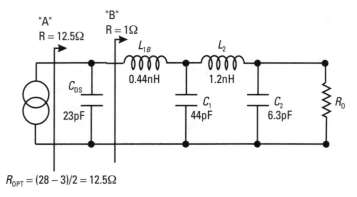

Figure 7.26 IMT matching network, 1.8–2.0 GHz.

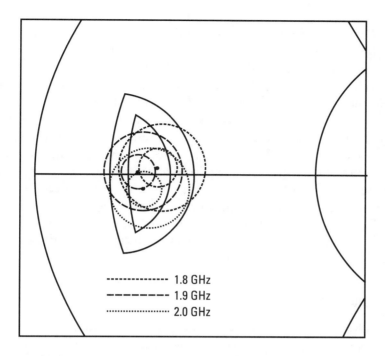

Figure 7.27 IMT matching sensitivity to imperfect 50-Ω termination, for 1.2:1 and 1.6:1 mismatch. The 1-dB and 2-dB power contours shown, based on 12.5-Ω optimum power, match at plane A in Figure 7.26.

somewhat idealized, but also represent a fairly moderate IMT example, in terms of frequency and power levels. The practical consequence of this analysis is that the microwave circuitry surrounding any IMT device has to be designed and executed with much greater care and precision than would normally be necessary for cascading lower power 50-Ohm devices.

This is not just a matter of etching microstrip lines to a tighter tolerance to ensure an accurate 50-Ohm impedance. The large package size of a typical high power IMT has to be carefully accommodated in order to maintain the necessary impedance through the "package-to-circuit" interface. It is usually highly recommended to follow the manufacturer's recommendations in this area, even to the extent of copying the test fixture used by the manufacturer in precise detail. Any changes in such physical parameters as board material, board thickness, microstrip line width, and the details of the mechanical recess into which the device is fixed, can cause changes in the device performance in comparison to the results obtained using the recommended test fixture.

Once the IMT has been successfully integrated into its new environment, there is an additional need for care in designing the rest of the microwave elements. Bias networks have to be electrically "invisible" over the RF bandwidth (Section 7.4.3), and any subsequent power combining circuitry has to maintain the necessary VSWR precision. There is also an issue of phase tracking if two or more devices are being power combined. The manufacturer's alignment process for an IMT will usually focus on meeting gain, power, and input VSWR specs. Phase tracking between units may not be guaranteed, and as such may require external compensation by the user. Unfortunately, if such phase trimming is done in an imperfect 50-Ω environment, the power and gain performance of the device will also change as the phase is adjusted. All of these difficulties have been known to greatly reduce the apparent saving of design and production time through the use of IMTs.

7.4.3   Biasing Issues for IMTs

Internally matched microwave transistors have a frustrating tendency to operate perfectly in a test fixture supplied by the manufacturer, but to show significant deviations when placed in a different, although nominally electrically identical, environment. Some of the reasons for this were discussed in the last section. The key word here is "nominal." In some cases the fixture environment may have some small but critical deviations from a perfect 50-Ω match which effectively form part of the matching network. Another potential contributor to this phenomenon is the necessary provision of bias networks. At higher gigahertz frequencies, the standard method for inserting bias into a 50-Ω environment is through a quarter-wave short circuited shunt stub (SCSS), usually with a short circuit provided by a capacitor, allowing the application of the bias voltage (Figure 7.28). The requirement for possibly several amps of supply current will typically necessitate the use of a moderate, rather than very high, characteristic impedance for the stub.

Figure 7.29 shows that a 50-Ω, $\lambda/4$ SCSS will "pull" the 50-Ω impedance environment to a significant degree, measured in terms of the load-pull results shown in Figure 7.27, for bandwidths exceeding about 10%. In a practical bias insertion network, the "short" will be realized using a suitable bypass capacitor. It is important to recognize that even small parasitics associated with the nominal capacitor value can seriously mistune the SCSS resonance; Figure 7.29 shows the effect of a 0.5 nH and 1 nH series inductance and a 20 pF bypass capacitor. Such values are typical for higher breakdown capacitors which need to be used in power stages. Such bias networks, in

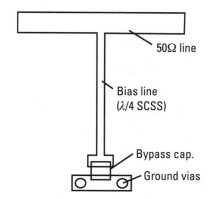

Figure 7.28 Typical IMT bias insertion network.

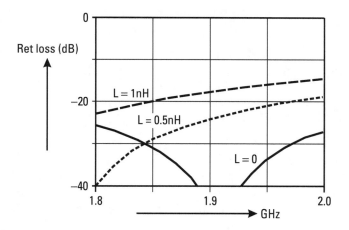

Figure 7.29 Effect of bypass capacitor parasitic on IMT bias SCSS.

critical applications such as these, should always be individually designed and the SCSS line length suitably adjusted. Once again, this underlines the importance of exact and complete duplication of the recommended electrical environment, even down to the type and manufacturer of the bypass capacitors. Much the same argument can be made for duplicating the position, size, type, and manufacturer of the necessary series blocking capacitors.

7.4.4 Power Combining of IMTs

It is common practice to combine the power of two or more IMTs. The subject of power combiners was covered in *RFPA* (Chapter 10), and the basics of the subject will not be repeated here. The move to higher frequencies and broader bandwidths does not alter the set of choices available to the designer in terms of combiner configurations, although the task of maintaining excellent match and low losses becomes progressively more challenging. In particular, the ubiquitous Wilkinson combiner will require two matching sections for bandwidths exceeding 10%, and losses from multiway "trees" escalate alarmingly. The problem of managing losses in higher power applications is clearly critical, not just from cost considerations but also from the increasing dissipation of power in the combining structure.

As discussed more briefly in *RFPA*, the basic rules in low-loss combiner design can be summarized as follows:

- Use of conductive elements having the largest possible cross-sectional area, (but compatible with the operating frequency);
- Elimination, or reduction, in the use of solid dielectrics;
- Use of enclosed transmission line structures;
- Care in placement and use of resistive elements;
- Meticulous compensation of input and output transition discontinuities.

A physical configuration which seems to satisfy these requirements, and which has been used extensively in critical higher frequency applications, is the so-called "suspended stripline" structure. This is, essentially, a way of realizing a stripline with an air dielectric, and is shown in Figure 7.30. The inner conductor is printed onto both sides of a thin, low-loss dielectric board. The upper and lower metallized lines are interconnected by a multitude of plated-through vias. The board is then sandwiched between metal plates which have cavities machined into them to form enclosed transmission lines. A typical Wilkinson-type combiner can be laid out much the same way as would be done using normal single layer microstrip, but the resulting structure will display much lower loss.

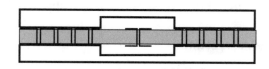

Figure 7.30 Cross section of "suspended stripline" transmission line.

The requirement for isolation resistors in a classical Wilkinson combiner poses some practical questions in higher frequency applications. The key issue is to ensure that such resistors are truly "nodal" and interconnect only between the combiner ports, without any ground coupling. In a typical open microstrip realization, such as that shown in Figure 7.31, the resistor will have significant physical dimensions and will inevitably couple to the ground plane. This effect can cause significant additional loss. At higher power levels the resistor size must be increased to allow for the possibility of greater heat dissipation, causing a further increase of this effect. One possibility to be considered in the most critical loss-sensitive applications is to dispense with the resistors. If the devices to be combined are accurately matched, the resistors have no function in normal operation; the odd-mode currents will be zero, and isolation between the devices is unnecessary. Experience seems to indicate that this is still something of a high-risk strategy unless some additional isolation between the separate power transistors can be provided. One such configuration would be to combine quadrature balanced pairs of devices. As discussed in *RFPA*, it seems a moot point as to why the Wilkinson structure, having such significant obstacles to low-loss realization, has remained the default choice for power combining applications. The quadrature hybrid coupler seems to have substantial benefits in terms of loss and bandwidth, and has much easier load resistor requirements. The need to move into more complex and expensive three-dimensional structures in order to reduce the insertion loss of the Wilkinson structure would seem to negate the simplicity advantage it has in less critical applications.

7.5 Distributed Amplifiers

The distributed amplifier (DA) has played a significant part in the military microwave era. Another vintage tube technique reincarnated [8], the DA proved to be one application where MMIC technology could outperform MIC hybrids and as such has received much emphasis in the literature and

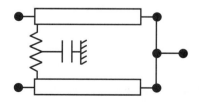

Figure 7.31 Ground plane capacitance effect on high power Wilkinson combiner.

the R&D labs over the last decade or two. Broadband DA MMICs, typically operating over several octaves of microwave bandwidth and delivering up to several hundreds of milliwatts of power, have been the most successful commercial microwave MMIC products. Electro-optic applications, which require a lower band limit extending down to dc, have found the DA to be the most satisfactory solution for high-speed laser drivers and detector video amplifiers. Microwave instruments have also made extensive use of commercial DA MMIC products.

The role of the DA in high-efficiency PA applications is more questionable. Unless there is a requirement for multioctave bandwidths, the efficiency and gain of a typical DA is found to be low. This is mainly due to inefficient use of transistor periphery, and is a fundamental problem with the DA configuration. It is therefore inappropriate to cover this subject in depth, but some indications of the problems associated with DA design in a power context are illustrated in a typical small signal DA design, shown in Figure 7.32.

The basic concept of a DA is to take a number of transistors and form artificial transmission lines using inductors connected between successive gates, on the input, and drains, on the output, as shown in Figure 7.32. This form of interconnection can neatly overcome the matching problem which an individual device would present over multioctave bandwidths. Each artificial line is designed to be approximately 50Ω and has a terminating resistor at the opposite end to the active signal port. A gain and return loss plot, shown in Figure 7.33, shows that a flat gain response, with reasonable match, can be obtained over many octaves of bandwidth. Such results, along with

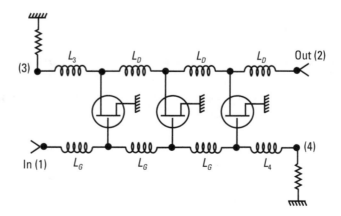

Figure 7.32 2–18-GHz distributed amplifier schematic. Ideal small signal FET model assumed ($c_{gs} = 0.25$ pF, $r_g = 5\Omega$, $g_m = 30$ mS, $c_{ds} = 0.075$ pF). Circuit values: $L_G = 0.4$ nH, $L_D = 0.4$ nH, $L_4 = 0.2$ nH, $L_5 = 0.6$ nH.

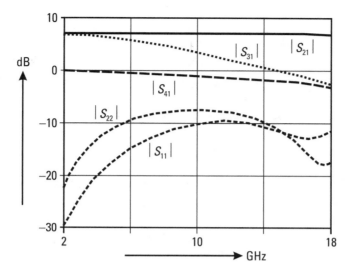

Figure 7.33 Distributed amplifier performance.

experimental verification, caught the attention of the military microwave community in the early 1980s and became the focus of much research and analysis [9–12].

Unfortunately, these appealing results do not stand up so well under closer scrutiny, particularly from a viewpoint of gain and power per unit device periphery. The gain and VSWR results, shown in Figure 7.33, do not tell the full story as to how such a circuit is really working from a power-management point of view. It is quite typical, for example, that the individual devices contribute unequal proportions of power to the output port, and reach compression conditions at different drive levels. There can also be a substantial wastage of power in the drain termination resistor; this can be seen in Figure 7.33, where the DA is analyzed as a terminated four-port. It has even been proposed that for power-sensitive applications the RF output can be extracted from both ends of the drain line using a suitable power combiner [12]. It should also be noted that in the example quoted, the overall gain is lower than that which could be obtained from a single device at any frequency in the DA bandwidth. The issue as to whether the achievement of a moderate broadband match merits the use of three or four times the number of transistors, for less gain and similar power overall than a single device, has been much debated. It can even be argued that many DA designs are really just paralleled transistors with the line terminations forming frequency-tailored lossy matching elements. At the low end of a decade band design, this would appear to be a more pragmatic description of the operation of a typical DA circuit.

The artificial transmission lines formed by the input and output interconnections will have a cutoff frequency which is a strong function of the gate and drain capacitance of the individual devices. This has the effect of limiting the upper frequency of a DA design for a given device. As the periphery of the device is increased in an attempt to obtain a higher power design, the upper frequency limit for multi-octave DA design is reduced. Various innovations have been demonstrated which can reduce the impact of the cutoff frequency; these include the use of series capacitors on the gates and the tapering of the individual transistor peripheries [10]. Additional pre-matching elements at each gate and drain can also improve the overall return loss and power distribution [11]. But the DA has largely been restricted to lower power applications, with reported power levels from uncombined DAs being below 30 dBm for Ku-band designs.

The success of the DA, despite these various perceived technical weaknesses, makes an interesting case study. Although originally developed for decade bandwidth applications, where there is truly no alternative, commercial DA products have found their way into many lower bandwidth microwave applications, often replacing MIC hybrids. They are, in effect, the broadband equivalent of the IMT discussed in the previous section. Ease of use, it seems, is a selling point which can still override the mainstream considerations of cost and design efficacy.

7.6 Conclusions

At higher gigahertz frequencies, power amplifier design becomes something of a poor relation to the highly efficient quasi-linear designs that can be realized below 2 GHz. This is due primarily to the limited speed of even the fastest available technology for making RF power transistors. Limited gain and substantial parasitics severely limit successful implementation of classical concepts such as Class AB. Broader bandwidths limit the ability to provide suitable harmonic terminations, and PA designers are frequently forced into using Class A operation, or something rather close to it.

Conventional PC board circuit design methods can be used with packaged transistors well beyond Ku band (12–18 GHz), but only for narrowband applications. Based on experience with satellite TV receivers it seems likely that future commercial applications will continue to stretch PC board circuit techniques. Broader bandwidths will, however, require either MMIC or MIC (chip-and-wire hybrid) circuit techniques, which could be adapted for high-volume use. Designers who are approaching broader bandwidth designs for the first time should make a point of at least reading some of the references quoted in this chapter [4, 5] on exact network synthesis methods, before resorting to the CAD optimizer.

High power, in the 100-W region, is limited by available transistor technology. Internally matched transistors above 3 GHz are mainly of the GaAs MESFET (or PHEMT) type, which are ultimately limited by low-voltage breakdown. Progress in Silicon Carbide (SiC) technology will probably not impact higher gigahertz applications, although Gallium Nitride (GaN) almost certainly will. The impact of HBT devices in lower power mobile applications seems to confirm the possible advantages of the bipolar transistor for applications requiring both linearity and high efficiency, discussed in Chapter 1 of this book. It remains to be seen whether higher power HBT devices will become commercially available. One of the continuous threads in the last 50 years or so of microwave technology development has been the phenomenon of technology being business-driven, rather than the other way around. A corollary of this is the lack of business interest in elegant or innovative technology. The power of digital electronics has started to make an impact on the traditionally analog-exclusive domain of RF engineering. The combination of business-driven technology and digital brute force would appear to offer the RF power system designer an interesting future.

References

[1]   Smith, C. R., et al., "The Microwave Power Module: A Versatile RF Building Block for High Power Transmitters," *Proc IEEE,* Vol. 87, No. 5, May 1999, pp. 717–737.

[2]   Brune, O., "Synthesis of a Two-Terminal Network Whose Driving-Point Impedance Is a Prescribed Function of Frequency," *J. Mathematics and Physics,* Vol. 10, No. 3, October 1931, pp. 191–236.

[3]   Bode, H. W., *Network Analysis and Feedback Amplifier Design,* New York: Van Nostrand, 1945.

[4]   Matthaei, G., L. Young, and E. Jones, *Microwave Filters, Impedance Matching Networks, and Coupling Structures,* New York: McGraw-Hill, 1964.

[5]   Minnis, B. J., *Designing Microwave Circuits by Exact Synthesis,* Norwood, MA: Artech House, 1996.

[6]   Lange, J., "Interdigitated Stripline Quadrature Hybrid," *IEEE Trans. on Microwave Theory and Technology,* Vol. MTT-17, No. 12, 1969, pp. 1150–1151.

[7]   Crescenzi, E. J., "Microwave Amplifier Manufacturing with Advanced Thin-Film MIC Technology," *Proc. IEEE Intern. Microw. Symp. MTTS-1989,* 1989, pp. 773–776.

[8]   Percival, W. S., "Thermionic Valve Circuits," British Patent 460562, January 1937.

[9]   Niclas, K. B., et al., "GaAs MESFET Distributed Amplifiers; Theory and Performance," *IEEE Trans. on Microwave Theory and Technology,* Vol. MTT-27, No. 6, 1983, pp. 447–456.

[10] Prasad, S. N., et al., "Power Bandwidth Considerations in Design of MESFET Distributed Amplifiers with Series Gate Capacitors," *IEEE Trans. on Microwave Theory and Technology,* Vol. MTT-36, No.7, 1988, pp. 1117–1123.

[11] Niclas, K. B., et al., "2–26.5 GHz Distributed Amplifier Using Declining Drain Line Lengths Concept," *IEEE Trans. on Microwave Theory and Technology,* Vol. 35, No. 4, 1986, pp. 427–435.

[12] Aitchison, C. S., et al., "The Dual Fed Distributed Amplifier," *Proc. IEEE Int. Microw. Symp.,* New York, Vol. 2, 1998, pp. 2911–2914.

Appendix

MESFET model used in SPICE simulations:

```
*generic GaAs MESFET; 1mm cell, Imax = 200mA
.model genfet gasfet level=2 beta=.310 vto=-2.5 alpha=2.0

*use model for internal cgs voltage dependency
* + b=3.0 lambda=.05 cds=.25pf cgs=.6pf cgd=.06pf rg=1

*set internal cgs, rg to small values to eliminate cgs
  variation
*rg, cgs now included in circuit file
 + b=3.0 lambda=.05 cds=.25pf cgs=.001pf cgd=.06pf
   rg=.001
* + b=3.0 lambda=.05 cds=.0001pf cgs=.0001pf cgd=.0001pf
   rg=.0001
```

Notes

The Spice "b" model typically has default settings which give an "aggressive" voltage dependency to the cgs capacitance. The above listing indicates a method for removing this dependency. Most of the simulations in this book use a 10-mm device, which requires a factor of 10 scaling applied to this model. This gives a device capable of delivering about 35 dBm in a loadline matched Class A design.

Selected Bibliography

The following books are recommended for further reading in related subject areas not fully covered in this book.

Cripps, S. C., *RF Power Amplifiers for Wireless Communications,* Norwood, MA: Artech House, 1999. Covers basic PA design at the circuit level and PA modes (referred to as *RFPA* throughout this book).

Harte, L., and S. Prokup, *Cellular and PCS/PCN Telephones and Systems,* APDG Publishing, 1996. An introduction and technical overview of modern cellular communications systems.

Kennington, P. B., *High Linearity RF Amplifier Design,* Norwood, MA: Artech House, 2000. A wide-ranging treatment of PA linearization topics.

Minnis, B. J., *Designing Microwave Circuits by Exact Synthesis,* Norwood, MA: Artech House, 1996. Detailed treatment of broadband microwave circuit synthesis methods.

Ojanperä, T., and R. Prasad, *Wideband CDMA for Third-Generation Mobile Communications,* Norwood, MA: Artech House, 1998. Introduction and comprehensive treatment of 3G communications systems.

Pothecary, N., *Feedforward Linear Power Amplifiers,* Norwood, MA: Artech House, 1999. Specific coverage of feedforward techniques, emphasis on practical system issues.

Glossary

ac Alternating current; term used generically to describe any sinusoidal signal

ACP Alternate channel power; term used more generally to describe the spectral distortion of spread spectrum signals, which appear as continuous bands

ADC Analog-to-digital converter

AGC Automatic gain control

AM Amplitude modulation

AM-AM Term used to describe gain compression in an amplifier, whereby a given increase in input signal results in a different change in the output level

AM-PM Amplitude modulation to phase modulation; a distortion process in a power amplifier, whereby increasing signal amplitude causes additional output phase shift

BER Bit error rate; a measure used to quantify the transmission integrity of a digital communications system

BJT Bipolar junction transistor

CAD   Computer-aided design

CDMA   Code division multiple access; a digital communications system which "chips" multiple signals in a pseudo-random fashion, making them appear like random noise unless a specific coding is applied

DAC   Digital-to-analog converter

DPA   Doherty power amplifier

DQPSK   Differential quadrature phase-shift keyed; a variation of QPSK

DSP   Digital signal processing

ECM   Electronic countermeasures; a generic term for military broadband microwave applications

EDGE   Enhanced Data rates for GSM Evolution; an evolution of the GSM system, providing data rates up to 384 Kbps

EER   Envelope Elimination and Restoration

EPA   Error power amplifier; used in feedforward loop

EPR   Ratio, usually expressed in decibels, of main PA power to EPA power; higher values imply higher system efficiency

EVM   Error vector magnitude; a more comprehensive measure of amplifier distortion which incorporates both gain compression and AM-PM

FET   Field effect transistor

GaAs   Gallium Arsenide; one member of a group of useful semiconductor materials which are compounds between group 3 and group 5 elements in the periodic table. These materials show higher mobility and higher saturation velocity than silicon and are therefore used in higher frequency applications.

GaN   Gallium Nitride; a newly emerging semiconductor material for high-power, high-frequency applications

GSM Global System for Mobilecommunication; most extensively used worldwide digital cellular network operating in the 900-MHz and 1,900-MHz bands

HBT Heterojunction bipolar transistor; most common variation of bipolar device used for higher frequency applications

HEMT High electron mobility transistor; an FET device in which a "sheet" of high-mobility material is created by the interaction between two epitaxial layers

IF Intermediate frequency

IM, IMD Intermodulation distortion

LDMOS Laterally diffused metal oxide semiconductor; derivative of silicon RF MOS technology for higher frequency applications

LINC Linear amplification using nonlinear components

LO Local oscillator

LUT Look-up table

MCPA Multicarrier power amplifier

MIC Microwave integrated circuit; a term used in the pre-MMIC/RFIC era to describe a miniaturized chip-and-wire microwave hybrid circuit technology

MMIC Microwave monolithic integrated circuit; a term now used, by convention, to describe monolithic integrated circuits which operate above 3 GHz

MOS Metal oxide semiconductor; usually silicon-based semiconductor technology

NADC North American Digital Cellular; original digital cellular system used in the United States

PAE Power-added efficiency; efficiency definition for an amplifier which accounts for the RF input drive

PEP Peak envelope power

PHEMT Pseudomorphic high electron mobility transistor; a higher frequency variant on the HEMT

PUF Power utilization factor; a term introduced in *RFPA* to define the "efficacy" of a PA design; defined as the ratio of power delivered in a given situation to the power delivered by the same device with the same supply voltage in Class A mode

QPSK Quadrature phase-shift keyed; a generic term for a variety of modulation systems which carry information only in the phase, not the amplitude, of the RF carrier

RFIC Radio frequency integrated circuit; a monolithically integrated device operating in the "RF" range, typically up to 3 GHz

RFPA Radio frequency power amplifier

RRC Raised root cosine; a filter of the Nyquist type, which allows a QPSK signal to be bandlimited without losing any modulation information

SCSS Short-circuited shunt stub; a distributed matching element used at microwave frequencies

SiC Silicon Carbide; a semiconductor material which offers high voltage and high velocity saturation but low mobility; potentially useful as a high power RF device technology in the low gigahertz frequency range

SPICE A general-purpose time domain nonlinear simulator, available in many implementations including shareware

SSB Single sideband; a variation on AM developed in the 1950s, in which only one modulation sideband of an AM signal is transmitted

WCDMA Wideband CDMA; a 3G system (e.g., "UTRA," or "UMTS"), based on CDMA, with data rates up to 2 Mbps

About the Author

Dr. Steve C. Cripps obtained his Ph.D. from Cambridge University in 1974. From 1974 to 1980, he worked for Plessey Research (now GECMM) on GaAs-FET device and microwave hybrid circuit development. He joined the solid-state division of Watkins-Johnson (WJ) in Palo Alto, California, in 1981, and since that time has held various engineering and management positions at WJ, Loral, and Celeritek. His technical activities during that period focused mainly on broadband solid-state power amplifier design for ECM applications. He has published several papers on microwave power amplifier design, including a design methodology that has been widely adopted in the industry.

Since 1990, Dr. Cripps has been an independent consultant, and his technical activities have shifted from military to commercial applications, which include MMIC power amplifer products for wireless communications. In 1996 he returned to England, where his focus is high-power linearized power amplifiers for cellular and satellite communications applications, and the characterization and modeling of high-power RF transistors.

Index

10-W PA designs, 143–44
  with different *k*-factor selections, 143
  phase response, 144
  See also Power amplifiers (PAs)

Adjacent channel power (ACP), 77
  asymmetry response, 77
  distortion, 81
Aluminum nitride (AIN) substrates, 282
AM-AM, 76
  characteristics, 84, 85
  compression characteristic, 83
  curves, 85
  distortion, 83, 84
  distortion measurement, 98
  dynamic, 95
  dynamic distortion plots, 100
  dynamic measurement test setup, 99
  fifth-degree, 85
  phase shift and, 95
  precision and, 88
Amplifiers
  auxiliary, 147–48
  balanced, 270–73
  BJT, 19
  Class A, 3, 12
  Class AB, 1–32
  Class B, 7, 9, 10

Class C, 33, 40, 42
distributed (DA), 293–96
feedback, 115
feedforward, 197–255
FET, 19
microwave power, 257–97
push-pull, 68
See also Power amplifiers (PAs); Radio
  frequency power amplifiers
  (RFPAs)
Amplitude
  error signal, 148
  generator voltage, 26
  IM3, 174
  IM, 90
  input voltage, 8
  output voltage, 8, 11
Amplitude envelope feedback, 121–36
  analysis schematic, 122
  attenuator characteristic at envelope
    domain, 127
  attenuator drive characteristic, 122
  bandwidth limitation, 123
  compensating delay line, 130
  delays, 127
  first-/third-order PA characteristic, 124
  as form of predistortion, 123
  limitations, 123

Amplitude envelope feedback (continued)
   linearization loop waveforms, 135
   modulation period, 129–30
   quasi-static response, 125
   RFPA characteristic, 127
   SPICE simulation, 131
   two-carrier excitation simulation, 134
   two-carrier IM3 response, 126
   See also Envelope feedback
AM-PM, 76, 81
   asymmetric, 95
   component at IM3 frequencies, 97
   contribution to IM level, 84
   correction in feedforward loop, 204–8
   correction loop, 138
   correction with envelope domain
      feedback, 137
   curves, 85, 86
   distortion, 83, 84
   distortion in main PA, 203
   distortion measurement, 98
   dynamic, 95
   dynamic distortion plots, 100
   dynamic measurement test setup, 99
   EPA power requirements and, 204
   fifth-degree, 86
   improved performance, 104
   lagging, 138
   leading, 138
   magnitudes, 85
   measurable process, 82
   peak amplitude, 96
   phase, 95
   phase angle, 96
   precision and, 88
   reduction of, 98
   removing/neutralizing, 84
   reversal of direction in, 85
   scaling factor, 83
   as secondary importance, 208
AM-PM effects, 90, 103, 205
   detrimental, 233
   ignoring, 229
   in main PA, 231–33
   on error vector magnitude, 207
   on feedforward loop correction
      signal, 206

Analog predistorters, 179–87
   categories, 179
   compound, 179, 187
   cuber as, 183
   mesa resistor as, 181, 182
   simple, 179
   See also Predistorters; Predistortion
Analog-to-digital converter (ADC), 149
Asymmetrical Doherty PA, 44–47
   benefits, 56
   current and voltage characteristics, 44
   defined, 44
   peaking function, 46
   See also Doherty PA (DPA)
Automatic gain control (AGC), 122
Auxiliary amplifiers, 147–48
   closed loop and, 148
   compensation power requirement, 213
   compression compensation, 211
   lowest, 211
   for restoring gain compression, 212
   voltage level requirement, 212

Balanced amplifiers, 270–73
   6-18 GHz medium-power
      schematic, 273
   benefits, 270–71
   illustrated, 271
   performance schematic, 272
   See also Broadband microwave power
      amplifiers
Bandpass filter
   transformed into matching
      network, 268
   two-section prototype transformed
      into, 267
Bandwidth
   amplitude envelope feedback loop, 123
   "real," 267
   video detection, 245
"Bazooka" structure, 279
Bessel functions, 96
Bias insertion networks, 285
Bipolar junction transistor (BJT), *xii*, 9
   amplifiers, 19
   base-emitter capacitance, 21
   Class AB. *See* BJT Class AB RFPA

Class A RFPA schematic, 20
device operation, 16
frequency analog circuit design, 19
gain and efficiency, 24
gain compression/efficiency vs. output
    power, 24
for high efficiency linear RFPA
    applications, 25
input impedance, 18
linearizing response to, 28
model illustration, 17
normalized transfer characteristic, 17
operation features, 16
RF model, 16–29
Si device, 21
Spice simulated waveforms, 20
thermal considerations, 18
transfer characteristics, 19, 23
BJT Class AB RFPA, 26–29
circuit, 22
current waveforms, 23, 27
design issues, 28–29
gain and efficiency, 27, 28
on-chip resistors, 29
schematic, 26
See also Bipolar junction transistor (BJT)
Broadband matching, 259–70
Broadband microwave power amplifiers
balanced, 270–73
Class AB operation, 273, 274
defined, 258
design, 259–79
design issues, 273–79
efficiency, 277
introduction, 259
load resistance, 276
matching with network
    synthesis, 259–70
peak-to-peak RF voltage, 277
push-pull schematic, 274
push-pull waveforms, 275, 278
See also Microwave power amplifiers
Broadband push-pull waveforms
Class AB, 278
illustrated, 275
Budget feedforward systems, 235, 253–55
amplifier chains, 255

defined, 254
illustrated, 255
simulation, 254
See also Feedforward loop; Feedforward
    systems
Butterworth response, 262

CAD optimizer, 270, 297
CAD simulation, 260, 262
Cancellation
Class AB compression, 30
errors, 244
IM3, 168
outphasing, 60
Cartesian Loop, 113, 119–20
defined, 119–20
linearization system, 119
Chebyshev filter, 262
parameters, 262
second-order, 265
Chebyshev lowpass prototype network
designing, 264
responses, 265
transmission function, 265
Chebyshev polynomials, 264
Chireix PA, 58–72
additional component, 58
analysis schematic, 61
combiner, 63
compensating reactances, 66
conclusions, 71–72
configuration illustration, 59
with conventional power
    combiner, 69–71
defined, 58
dependencies, 59–61
discussion, analysis, simulation, 62–69
efficiency, 67–68
introduction and formulation, 58–62
load-pulling effect, 60
outphasing cancellation, 60
outphasing circuit schematic
    simulation, 65
outphasing PA simulation results, 67
outphasing shift, 62
outphasing technique, 58

Chireix PA (continued)
  output matching and balun
      realization, 69
  phase shifters, 59–60
  power/efficiency plots, 67
  saturated amplifier assumption, 59
  shunt reactance, 64, 65
  simulation of Class FD, 63, 64
  variations, 69–71
  See also Power amplifiers (PAs)
Class A amplifiers
  bias point, 3
  BJT schematic, 20
  linear characteristic, 12
Class AB amplifiers, 1–32
  analysis, 10
  BJT, 22, 26–29
  broadband, efficiency/PBO
      performance, 278
  broadband push-pull waveforms, 278
  classical, 2–9
  compression cancellation, 30
  deep, 102
  defined, 1
  efficiency, 5, 8, 9, 273
  gain characteristics, 6
  key circuit element, 4
  linearity, 4
  linearity "zone," 15
  output current, 11
  output power, 8
  PAs, 21
  "quiescent" current setting, 5
  schematic, 3
  tunnel vision, 1
  in "underdrive" case, 7
  waveforms, 3
Class B amplifiers
  classical, 9
  operation, 7
  quiescent bias point, 10
  theoretical linearity, 10
Class C amplifiers, 33, 40
  operation, 42
  peaking device, 41
Classical Class AB modes, 2–9
Classical Doherty configuration, 37–42

amplitudes, 41
bias adaptation scheme, 50
current and voltage, 39
efficiency, 41
fundamental current component, 40
ideal device characteristics, 37
implementation stumbling blocks, 41
peaking PA realization, 50
RF current, 41
voltage amplitude, 38
See also Doherty PA (DPA)
Coaxial balun structure, 279
Code division multiple access
      (CDMA), 9, 46
Compensating delay line, 130
Composite PD/PA response
  IM3, 160, 168
  with PD having third-/fifth-degree
      characteristics, 161, 169
  with PD having unmatched expansion
      characteristics, 162
  with third-degree PD, 160, 167
Compound analog predistorter, 179
  cuber, 187
  process, 187
  See also Predistorters
Compression
  AM-AM characteristic, 83
  auxiliary PA, 211, 212
  Class AB amplifier cancellation, 30
  gain, 160, 199
  relative power requirement for
      restoring, 213
Compression adjustment, 218, 224–28
  defined, 218
  FFW loop as detraction, 226
  FFW loop distortion, 225
  gain, 233
  importance, 224
  simulation of, 234
  See also Feedforward loop
Coplanar waveguide, 284
Couplers
  directional, 208–11
  error insertion, 208–16
  Lange, 285
  microstrip, 285

microwave, 208
Coupling factor, 218
Cube-law device characteristics, 11, 12
Cuber
    compound, 187
    configuration, 181, 182
    defined, 183
    input to, 183
    measured performance of, 185
    nonlinear elements in, 183
    as predistorter, 183
    using, 183

Delay
    amplitude envelope feedback
        system, 127
    closed-loop, 121
    envelope feedback loop, 119
    group, 120
    high-power PA, reduction, 139
    inserting, 127
    in multicarrier 3G applications, 148
    in multistage high power PA
        assemblies, 145
    PA, reduction, 121
    RFPA, 113
    RFPA gain stages, 144
    vector envelope feedback, 139
De Moivre's theorem, 91
Diamond heatsinks, 282–83
Dicke receiver, 246
    calibration source, 247
    illustrated, 247
Differential quadrature phase shift keyed
        (DQPSK) format, 106
Digital signal processor (DSP), 33
    algorithmically-based correction
        system, 193
    algorithmic precision for, 177
    calibration system, 193
    computation process, 193
    control elements, 189
    controllers, 112
    phase control, 72
    speed and availability, 194
    techniques, xii
    See also DSP predistortion

Digital-to-analog converter (DAC), 189
Diode PDs, 162
Direct feedback, 114
Directional couplers, 208–11
    coupling coefficient, 209
    in microstrip MIC, 284
    "misconception," 210
    as signal combiner, 211
    with single sinusoidal signal
        excitation, 209
    transmission coefficient, 209
    transmission factor restoration, 214–15
    with two sinusoidal cophased input
        signals, 210
Distortion, 75
    ACP, 81
    AM-AM, 83, 84, 98, 100
    AM-PM, 83, 84, 98, 100
    close-to-carrier IM, 81
    compression adjusted FFW loop, 225
    feedforward-enhanced power
        combiner, 253
    FFW loop, 219, 223
    production isolation, 202–3
    third-degree, 83, 165
Distributed amplifiers (DAs), 293–96
    broadband, MMICs, 294
    concept, 294
    in high-efficiency PA applications, 294
    multi-octave design, 296
    performance, 295
    role, 293
    success, 296
Dogleg characteristic, 14
Doherty-Lite, 47–49
    backoff efficiency, 47
    benefits, 47–48
    bias settings, 48
    defined, 47
    efficiency improvement, 48
    main and peaking functions, 47
    simulation, 49
Doherty PA (DPA), 34–57, 241
    amplitudes, 41
    analysis, 42
    asymmetrical, 44–47
    Class A, efficiency curves, 52

Doherty PA (DPA) (continued)
    classical configuration, 37–42
    conclusions, 56–57
    Doherty-Lite, 47–49
    efficiency, 41
    FFW loop using, 241
    ideal, 42, 43
    ideal device characteristics, 37
    ideal harmonic shorts, 35
    idealizations used in analysis, 36
    I_{max} values, 36–37
    impedance converter, 34
    impedance requirement, 54
    implementation stumbling blocks, 41
    introduction and formulation, 34–37
    linearity, 57
    main device impedance load, 54
    matching technologies, 52–56
    multiple, 56
    peaking amplifier configuration, 49–52
    practical realization schematic, 53
    RF current, 41
    simulation of matched version, 55
    simulation with two GaAs MESFET
        devices, 51
    two-device, schematic, 35
    variations on classical
        configuration, 42–49
    See also Power amplifiers (PAs)
Doherty with nonlinear peaking
        device, 42, 43
    amplitudes, 43
    efficiency characteristics, 43
    RF current, 43
Double feedforward loop, 249–52
    benefits, 250
    defined, 249–50
    EPA2, 250–51
    illustrated, 251
    logic, 250
    unpopularity, 250
    See also Feedforward loop
Drift
    compensation scheme
        implementation, 245
    compensation scheme
        requirements, 244

    domain, 245
    as enemy, 244
    reducing, 246
    slowness, 246
    test, 243
DSP predistortion, 187–94
    with algorithmic process, 193
    LUT-based, 192
    scheme illustration, 189
    See also Digital signal processor (DSP);
        Predistorters; Predistortion

Efficacy, 236
Efficiency
    BJT RFPA, 25, 27, 28
    broadband microwave power
        amplifiers, 277
    Chireix PA, 67–68
    Chireix PA with power combiner, 70
    Class AB amplifier, 5, 8, 9
    cube-law device, 12
    Doherty-Lite, 47, 48
    Doherty PA, 41
    Doherty using nonlinear peaking
        device, 43
    even harmonic enhancement, 14
    FFW loop, 236–41
    ideal Doherty, 43
    linear high, 9
Electronic countermeasure (ECM)
        receiver, 259
Envelope detectors, 246
Envelope Elimination and Restoration
        (EER) method, 33, 34
Envelope feedback, 117–19
    amplitude, 121–36
    AM-PM correction using, 137
    with auxiliary PA, 147
    as basis for LUT calibration, 150
    compensating delay line, 130
    defined, 117
    delay, 119
    drift domain, 245
    higher RF frequencies and, 118
    limitations, 118
    for LUT calibration, 192
    with output power control, 146

system block diagram, 117
techniques, 117
variations, 146–50
vector, 137–40
Envelope input sensing, 191–92
Envelope simulation, 73, 76
EPA2, 250–51
electrical length, 250
input signal components, 250
power level, 250, 251
power requirements, 251
See also Double feedforward loop; Error
power amplifier (EPA)
Equal-power splitters, 185
Equal-ripple filter, 263
Error insertion coupler, 208–16
application, 211
directional, 209–11
transmission factor, 216
transmission loss, 215, 228
Error PA ratio (EPR), 217
defined, 217
inner loop, 251
PBO tradeoff, 239
requirement, 236
Error power amplifier (EPA), 200
correction, redrawn, 207
correction signal, 205
design, 237–38
distortion products, 201
EPA2, 250–51
FFW loop simulation change, 233–35
FFW loop simulation output, 231
"flea-power," 235
gain, 202
nonlinearity, 201
power, as quantitative measure, 204
power rating, 229
power requirement, 201, 204, 205
power selection, 202
power specification, 239
required power capability, 202
required power output, 203
for restoring coupler transmission
factor, 214
Error vector magnitude (EVM), 77, 84
AM-PM reflect on, 207

concept, 107
measurement, 108
specification, 107, 108
Fast Fourier transform (FFT), 77
Feedback
amplitude envelope, 121–36
classical amplifier configuration, 115
compensation in drift domain, 245
direct, 114
envelope, 117–19
gain equation, 115–16
indirect, 111–12
introduction to, 111–14
linearization effect, degradation of, 116
low latency PA design, 140–46
negative, 111
"rule of thumb," 126
techniques, 111–51
vector envelope, 137–40
Feedforward-enhanced power
combiner, 252–53
cost, 252
distortion, 253
illustrated, 252
performance, 253
Feedforward loop, 198–203
as additive process, 198
AM-PM correction in, 204–8
AM-PM effect on, 206
analysis illustration, 217
basic action, 198
budget, 254
cancellation errors, 244
closing, 241–49
distortion, 223
with Doherty PA, 241
double, 249–52
with drift domain envelope
feedback, 245
efficiency, 236–41
efficiency plot, 239
error signal, 219
gain and phase tracking system, 247
illustrated, 199
IM3 performance, 224, 225, 226
IMs, 233

Feedforward loop (continued)
  multicarrier response, 236
  "normalized adjustment," 218,
     219, 226
  operation, 199
  output distortion, 219
  power drain, 237
  third-degree analysis, 216–29
  tracking error effect on, 223
  two-carrier IM3 response, 222
Feedforward loop simulation, 229–36
  AM-PM effects and, 229
  compression adjustment, 234
  effect of AM-PM in main PA, 231–33
  EPA output, 231
  EPR change, 233–35
  gain compression adjustment, 233
  gain/phase tracking, 235
  multicarrier simulation, 235–36
Feedforward systems, 148, 153, 197–255
  benefits, 242
  budget, 235, 253–55
  with built-in "virtual" bench test, 248
  "compression adjustment," 218
  conclusions, 255
  correction signal, 241–42
  defined, 197
  double, 249–52
  drift in, 244
  efficiency, 237
  enhanced power combiner, 252–53
  for envelope time domain
     correction, 245
  EPA power consumption, 237
  equipment, 242
  error insertion coupling, 208–16
  introduction, 197–98
  measurement, 242–43
  PA gain/phase response changes, 242
  PBO amount, 237
  performance, 223
  response time, 242
  setup, 242
  variations, 249–55
Field effect transistor (FET), 2, 9
  adjacent channel power (ACP)
     responses, 29

  amplifiers, 19
  approximation to dogleg
     characteristic, 14
  characteristic with gain expansion, 30
  device operation, 16
  intermodulation (IM) responses, 29
Filter synthesis theory, 262
Four-carrier third-order IM spectrum, 176

GaAs MESFET, 66
  Doherty PA simulation with, 51
  impact, 259
  phase angle bias dependency, 104
Gain
  BJT RFPA, 27, 28
  Class AB characteristics, 6
  EPA, 202
  feedback equation, 115–16
  FFW loop simulation, 235
  ninth-degree, 93
  nonlinearity in, 30
  PA output stage, 6
  predistorter, 156, 157
  reduction, at low drive levels, 30
  video, 139
Gain compression
  adjustment, 233
  composite characteristics, 160
  third-degree, 199
GSM EDGE signals, 107
  constellation illustration, 108
  EVM specification, 107

Harmonic efficiency enhancement, 14
Heatsinks, 282–83
Heterojunction bipolar transistor
     (HBT), *xii*
  external harmonic circuitry, 21
  handset PAs, 25
  impact in lower power applications, 297

Ideal Doherty, 42, 43
  with Class C peaker, 42
  device characteristics, 37
  harmonic shorts, 35
  See also Doherty PA (DPA)
Impedance
  converter, 34

RF transistor, 141
transformation, 267, 268
Indirect feedback techniques, 111–12
Inductors
approximation, 270
network transformations, 261
Intermodulation (IM)
amplitudes, 90
asymmetry in RFPAs, 94–105
close-to-carrier distortion, 81
fifth-order characteristics, 89
mid-regime correction, 126
PBO curves, 78
PBO sweeps, 79
phase measurements, 90
plots, 88
seventh-order characteristics, 89
two-carrier response, 80
upper/lower sideband asymmetry, 81
See also Third-order intermodulation
(IM3)
Internally matched microwave transistors
(IMTs), 287–91
biasing issues, 290–91
bias insertion network, 291
bias SCSS, 291
high power, 289
integration, 290
matching issues, 287–90
matching network illustration, 288
matching sensitivity, 289
power combining of, 291–93
uses, 287
Irreducible cubic, 125

Khan restoration loop, 120
"Knee" value, 2

Lange coupler, 285
Latency
high Q-factors and, 113
PA, 130
RFPA, 113
Laterally Diffused Metal Oxide
Semiconductor (LDMOS), 66,
92, 237
Loadline theory, 288

Load resistors, 8
Look-up tables (LUTs)
calibration, 150
DSP drive from, 193
dynamic refreshing system, 191
envelope feedback for calibration, 192
loading-with dynamic calibration
signal, 192
longevity, 191
precision, 190–91
predistorter use of, 188
Low latency PA design, 140–46
Lowpass filters, 133
Lowpass matching network, 55, 141

Matched PD, 159
Matching network
bandpass filter transformed into, 268
four-element, 269
IMT, 288
lowpass, 55, 141
synthesis procedure, 270
Memory, in RFPAs, 94–105
Mesa resistor, 181, 182
MESFET
GaAs, 51, 66, 104, 259
model in SPICE simulation, 299
Metal-insulator-metal (MIM)
capacitors, 286
Microstrip couplers, 285
Microstrip MIC, 283, 284
directional coupler, 284
transmission line illustration, 284
Microwave couplers, 208
Microwave Integrated Circuits
(MICs), 280–86
advanced processes, 286
components and structures, 283–86
configuration for higher dissipation
components, 282
defined, 280
disadvantages, 281
elements, 280
illustrated, 280
"lumped" elements, 286
material properties, 283
modules, 281

Microwave Integrated Circuits (MICs)
   (continued)
   substrate and heatsink
      materials, 281–83
   technology, 258, 283
   transmission lines, 284
Microwave power amplifiers, 257–97
   broadband design, 259–79
   conclusions, 296–97
   design with prematched
      modules, 287–93
   distributed, 293–96
   introduction, 257–59
   microwave circuits/MIC
      techniques, 280–86
   technology eras, 257–58
Military microwaves, 258
Modern multicarrier power amplifier
   (MCPA)
   era, 49
   feedforward system for, 255
"Modulation domain" frequency, 221
Modulation period, 129–30
Monolithic microwave IC (MMIC)
   DA, 294
   development, 286
Multicarrier PA spectral response, 170
Multicarrier PD/PA spectral
      response, 171–73
   matched third-/fifth-degree PD, 172
   notcher PD, 173
   third-degree only, 171
Multiple Doherty PA, 56
   flatter efficiency PBO characteristic, 56
   principle, 56
   schematic, 57
   See also Doherty PA (DPA)
Multistage RFPAs, 145

N=2 bandpass filter, 268
Negative feedback, 111
Network synthesis, 259–70
Neutralization process, 36
Nonlinearities
   attenuator, 147
   EPA, 201
   gain, 30

PA, 73–110
"Normalized adjustment," 218, 219, 226
North American Digital Cellular (NADC)
   signals, 106
   constellation illustration, 107
   peak-to-average ratio, 107
   phaseplane trajectory, 106
Norton transformation, 268
"Notcher" predistortion, 172–76
N-section lowpass prototype filter, 264
Nyquist raised root cosine (RRC)
   filter, 106

Open-circuited shunt stubs (OCSSs), 270
Organization, this book, xii–xiii
Oscillation, 115, 133
Outphasing
   for AM signal construction, 69
   cancellation, 60
   circuit simulation schematic, 65
   with conventional power combiner, 70
   defined, 58
   impedance shift, 62
   shunt reactance effect and, 65
   simulation results, 67
   See also Chireix PA
Output power
   Class AB amplifier, 8
   envelope feedback and, 146
   gain compression/efficiency vs., 24

Packaged discrete components, 145
PD/PA
   analysis configuration, 163
   characteristic, 154
   spectral response, 171
Peak envelope power (PEP), 240
Peak power, 74
Peak-to-average ratios, 105–9
   GSM EDGE signal, 107
   NADC signals, 107
   problem, 106
   WCDMA, 105
Peak-to-peak RF voltage, 277
Periphery scale-up (PSU), 240
Phase
   control loop, 148
   detector, 139

linearity, 243
offset, 138, 243
tracking, 235, 247
transmission, of networks, 263
trimmer, 243
Pilot carrier tracking systems, 248
PIN diode, 180
Polar Loop system
defined, 120
illustrated, 119
Polynomial curve-fitting routines, 87
Power amplifier (PA)
    nonlinearities, 73–110, 129
conclusions, 109–10
envelope feedback system, 132
IM asymmetry, 94–105
introduction, 73–74
inverted, 186
peak-to-average ratios, 105–9
polynomials, 77–89
power series, 77–89
two-carrier characterization, 89–94
Volterra series, 77–89
Power amplifiers (PAs)
10-W designs, 143–44
auxiliary, 147, 211, 212
Chireix outphasing, 58–72
delay reduction, 121
design of 10W with different *k*-factor
    selections, 143
design tradeoffs, 144
design using prematched
    modules, 287–93
device technologies, 113
Doherty, 34–57
drift test, 243
feedback linearization, 114
latency, 130
low latency design, 140–46
microwave, 113
modeler advantages, 87
multicarrier, 170
peaking, 42, 49–52
peak-to-average ratios and, 105–9
push-pull, 274
with range of cutoff/conduction
    angles, 40

third-degree, 156–63
unpredistorted sweeps, 171
Power backoff (PBO)
9:1 rates, 230, 239
efficiency, 70
EPR tradeoff, 239
IM curves, 78
low levels, 188
sweeps of IM, 79
Power-combined modules, 149
Power combiner
Chireix PA with, 69–71
feedforward-enhanced, 252–53
IMT, 291–93
low-loss design, 292
outphasing with, 70
Power control, envelope feedback with, 146
Power series
coefficients, 93
composite PD/PA, 159
odd-degree, 91
PD, 163
for synthesis of PD function, 158
Power transistors, 280
Predistorters
analog, 179–87
basic action, 155
compound, 179, 187
cuber as, 183
design, 154
diode, 163
fifth-degree coefficients, 166
gain expansion, 156, 157
input signal to, 164
LUT use, 188
practical realization, 177–78
RFIC, 181
signal emerging from, 155, 164
simple, 179–80
simplicity, 154
third-degree, 156
third-degree coefficients, 165
Predistortion
amplitude envelope feedback as form
    of, 123
categories, 178
classes of practical realization, 162

Predistortion (continued)
    conclusion, 194–95
    defined, 153
    DSP, 187–94
    effective, 155
    for general PA model, 163–76
    ideal characteristic, 156
    introduction, 153–56
    matched, 159
    matched third-degree
        characteristic, 159–60
    matched third-/fifth-degree
        characteristic, 160–61, 168–69,
        171–72
    matched to third-degree
        only, 167–68, 170–71
    "notcher," 172–76
    performance, 154
    power series, 163
    techniques, 153–95
    third-degree PA, 156–63
    unmatched, 169–70
    unmatched third-degree gain expansion
        characteristic, 161–63
Prematched modules, 287–93
    biasing issues, 290–91
    introduction, 287
    issues, 287–90
    power combining, 291–93
Pseudomorphic high electron mobility
        transistor (PHEMT), 25
PUF ratio, 277
Push-pull amplifiers, 68, 274

Q-factor, 21, 142, 148
    for high-power devices, 140, 145
    latency and, 113
    for matching networks, 140
Quarter-wave short circuit shunt stub
        (SCSS), 66, 68, 290

Radio frequency integrated circuit (RFIC)
    alternatives, 286
    designers, 40
    predistorters, 181
Radio frequency power amplifiers (RFPAs)
    BJT Class A, 20
    BJT Class AB, 22, 23, 26–29

Class C, 33
delay, 113
design, 1
IM asymmetry in, 94–105
memory in, 94–105
multistage, 145
phase linearity, 243
push-pull, 274
"sweet spots," 6
transistors, 1
    See also Power amplifiers (PAs)
RF bipolars, 14, 16–29
RF Power Amplifiers for Wireless
        Communications, xi–xii
RF spectral domain, 76
RF time domain, 75

"Satcom" applications, 258
Schottky diodes, 184
Seidel system, 247
Series capacitors, 270
Short circuit shunt stub (SCSS)
    IMT bias, 291
    line length, 291
    quarter-wave, 66, 68, 290
    resonance mistuning, 290
Shunt diode limiter, 184
Signal combiner, directional coupler
        as, 211
Silicon Germanium (SiGe) technology, xii
Simple analog predistorters, 179–80
    advantages, 179–80
    defined, 179
    illustrated, 179
    limitations, 180
    typical performance, 180
    See also Predistorters
Single sideband (SSB) era, 33
SPICE simulation
    amplitude envelope feedback
        system, 131
    BJT Class A RFPA, 20
    MESFET model used in, 299
    PA and input controller, 133
Square-law detection, 246
Square-law device characteristics, 11, 12
Supply rail modulation effect, 102

Surface-mount (SMT)
    components, 181
    Schottky diodes, 184
"Suspended stripline" transmission line, 292

Third-degree FFW loop analysis, 216–29
    compression adjustment, 224–28
    conclusion, 228–29
    formulation and analysis, 216–22
    illustrated, 217
    quantification benefit, 216
    results summary, 228–29
    system, 216–17
    tracking errors, 222–24
    See also Feedforward loop
Third-degree nonlinearity, 92
Third-degree PA, 156–63
    composite PD/PA response, 160
    nonlinear, 159–63
    PD with matched third-degree
        characteristic, 159–60
    PD with matched third-/fifth-degree
        characteristic, 160–61
    PD with unmatched third-degree gain
        expansion characteristic, 161–63
    polynomial expression, 159
    two-carrier IMD responses, 159
Third-order intermodulation (IM3)
    amplitude, 174
    cancellation, 168
    combined higher sideband, 97–98
    combined lower sideband, 98
    components, 79
    FFW loop performance, 224, 225, 226
    frequencies, 81
    generation, 78
    of ideal transconductive device, 10
    in-band products, 175
    notching, 31
    plots, 88
    products, 174, 175
    products, nulling, 31
    sidebands, 174, 176
    spectral regrowth frequency band, 175
    two-carrier response, 126, 222
    two-tone products, 77
    See also Intermodulation (IM)

Time domains, 74, 75
    measurement, 75
    RF, 75
Time trajectory, 74
Total power receiver, 246
    defined, 246
    illustrated, 247
Tracking
    phase, 235, 247
    pilot carrier, 248
Tracking errors, 222–24
    effect of, 235
    FFW loop simulation, 235
Transistors
    bipolar junction (BJT), *xii*, 9, 16–29
    characteristics, 10
    field effect (FET), 2, 14, 16, 19
    heterojunction bipolar
        (HBT), *xii*, 21, 25, 297
    impedances, 141
    internally matched microwave
        (IMTs), 287–91
    limited switching speed of, 13
    power, 280
Traveling wave tubes (TWTs), 259
Two-carrier characterization, 89–94
    advantages, 94
    defined, 89
    dynamic envelope measurements, 101
    IM3 response, 222
    modeling procedure, 93–94
Two-section lowpass prototype
        filter, 265, 267

"Underdrive" concept, 7
Unmatched PD, 169–70

Vector envelope feedback, 137–40
    delay, 139
    gain block, 140
    phase detector, 139
    video gain, 139
    See also Envelope feedback
Video detection bandwidth, 245
Video gain, 139
"Virtual" bench test, 248
Volterra coefficients, 89
    derivation of, 91

Volterra coefficients (continued)
    normalized, 231
Volterra formulation, 73
Volterra phase angle, 180, 203
Volterra series, 82–85
    fifth-degree, 177
    inverted PA, 166
    nonlinear PA with, 163
    PA characteristics modeled with, 85
    phase angles, 82, 92
VSWR
    dependent ripple, 273
    interactions, 271
    levels, 288
    precision, 290
    response, 271, 273

Waveforms
    BJT Class AB, 23
    Class AB, 3
    envelope linearization loop, 135
    "maximally flat" even harmonic
        components, 13
    peak-to-peak swing, 13
    push-pull, 275
    simulated envelope, 133
    Spice simulated, 20
Wideband CDMA (WCDMA), 46
    peak-to-average ratios, 105
    signal magnitude trajectories, 74
Wilkinson combiner, 292–93
    ground plane capacitance effect on, 293
    isolation resistors in, 293

"Zero bias" operation, 9

Recent Titles in the Artech House Microwave Library

Advanced Techniques in RF Power Amplifier Design, Steve C. Cripps

Behavioral Modeling of Nonlinear RF and Microwave Devices, Thomas R. Turlington

Computer-Aided Analysis of Nonlinear Microwave Circuits, Paulo J. C. Rodrigues

Design of FET Frequency Multipliers and Harmonic Oscillators, Edmar Camargo

Design of RF and Microwave Amplifiers and Oscillators, Pieter L. D. Abrie

EMPLAN: Electromagnetic Analysis of Printed Structures in Planarly Layered Media, Software and User's Manual, Noyan Kinayman and M. I. Aksun

Feedforward Linear Power Amplifiers, Nick Pothecary

Generalized Filter Design by Computer Optimization, Djuradj Budimir

High-Linearity RF Amplifier Design, Peter B. Kenington

Introduction to Microelectromechanical (MEM) Microwave Systems, Hector J. De Los Santos

Microwave Engineers' Handbook, Two Volumes, Theodore Saad, editor

Microwave Filters, Impedance-Matching Networks, and Coupling Structures, George L. Matthaei, Leo Young, and E.M.T. Jones

Microwave Materials and Fabrication Techniques, Third Edition, Thomas S. Laverghetta

Microwave Mixers, Second Edition, Stephen Maas

Microwave Radio Transmission Design Guide, Trevor Manning

Microwaves and Wireless Simplified, Thomas S. Laverghetta

Neural Networks for RF and Microwave Design, Q. J. Zhang and K. C. Gupta

QMATCH: Lumped-Element Impedance Matching, Software and User's Guide, Pieter L. D. Abrie

RF Design Guide: Systems, Circuits, and Equations, Peter Vizmuller

RF Measurements of Die and Packages, Scott A. Wartenberg

The RF and Microwave Circuit Design Handbook, Stephen A. Maas

RF and Microwave Coupled-Line Circuits, Rajesh Mongia, Inder Bahl, and Prakash Bhartia

RF Power Amplifiers for Wireless Communications, Steve C. Cripps

RF Systems, Components, and Circuits Handbook, Ferril Losee

TRAVIS 2.0: Transmission Line Visualization Software and User's Guide, Version 2.0, Robert G. Kaires and Barton T. Hickman

Understanding Microwave Heating Cavities, Tse V. Chow Ting Chan and Howard C. Reader

For further information on these and other Artech House titles, including previously considered out-of-print books now available through our In-Print-Forever® (IPF®) program, contact:

Artech House
685 Canton Street
Norwood, MA 02062
Phone: 781-769-9750
Fax: 781-769-6334
e-mail: artech@artechhouse.com

Artech House
46 Gillingham Street
London SW1V 1AH UK
Phone: +44 (0)20 7596-8750
Fax: +44 (0)20 7630 0166
e-mail: artech-uk@artechhouse.com

Find us on the World Wide Web at:
www.artechhouse.com